D1559723

INDUSTRIAL WASTEWATER TREATMENT

A Guidebook

Joseph D. Edwards, Ph.D., P.E.

Library of Congress Cataloging-in-Publication Data

Edwards, Joseph D.
Industrial wastewater treatment: a guidebook / Joseph D. Edwards.
p. cm.
Includes bibliographical references and index.
ISBN 1-56670-112-0
1. Factory and trade waste—Purification. 2. Sewage—Purification. I. Title.
TD897.5.E33 1995
628.3—dc20 95-5800
CIP

Direct all inquiries to CRC Press, Inc., 2000 Corporate Blvd., N.W., Boca Raton, Florida 33431.

Lewis Publishers is an imprint of CRC Press

International Standard Book Number 1-87371-112-0
Library of Congress Card Number 95-5800
Printed in the United States of America 1 2 3 4 5 6 7 8 9 0
Printed on acid-free paper

THE AUTHOR

Joseph D. Edwards, P.E., Ph.D., has worked in industry, consulting, academia, and government. He has implemented new industrial processes, fixed old ones, designed and started up waste treatment processes, determined the cause of treatment plant upsets, assessed the impact of new production processes on waste management activities, and developed hazardous waste reduction plans. In addition, he has developed local limits for industrial discharges, reviewed process engineering reports for wastewater treatment systems, written permits for industrial dischargers, developed compliance plans, conducted analysis of the effects of discharges on municipal treatment plants, and helped to revise state regulations regarding the discharge of dangerous waste to the sanitary sewer. He is the co-inventor of a patented process for removing toxic organics and metals from manufacturing wastewater.

Dr. Edwards earned a B.S. in Chemistry, a M.S. in Civil Engineering, Air Resources, and a Ph.D. in Civil Engineering, Environmental Science and Technology, from the University of Washington. His Ph.D. dissertation evaluates the use of diffusion dialysis for the recovery of acids from spent process solutions. He is a registered Professional Engineer in the State of Washington. Dr. Edwards can be reached via e-mail at cleanh2O@halcyon.com.

ACKNOWLEDGMENTS

Thanks to the three shops who let me share their experience in the appendices and pilot test examples and who reviewed the text to help to ensure that it is accurate. Thanks to the shop that provided the photograph of the sink for the front cover.

Thanks to Rick Renaud, P.E. (Municipality of Metropolitan King County, Metro) and Professor Mark Benjamin (University of Washington) for supporting this book in its proposal stage and having confidence that it could be done.

Most of all I want to thank my wife, Belle Randall Edwards, for her help and encouragement. She read the book cover-to-cover and made many helpful comments. She also provided much needed moral support.

CONTENTS

DEDICATION

To my mother, Genevieve L. Edwards and the memory of my father, John M. Edwards.

1 HOW TO USE THIS BOOK

INTRODUCTION

Many businesses generate wastewater. Sometimes the wastewater must be treated before discharge to the sewer. How do you go about deciding if you need a wastewater treatment system? If you do need one, how do you put it in? There is more to wastewater treatment than installing a piece of equipment. It is important to understand if and why a treatment system is needed, and to be aware of your options, in order to select the type of system most compatible with your shop's operations.

This book is intended to help those who are faced with putting in a wastewater treatment system and to explain the methodology to others concerned with industrial pretreatment. The approach presented in this book applies to the design and implementation of a treatment system of any size. Organized as a step-by-step guide, the chapter outline presented below provides a brief overview of the steps involved, from start to finish. Gloss over it now, and refer to it later, if you ever feel overwhelmed or uncertain about what comes next. Each of the chapters that follows amplifies one of the steps in the process. You may not need to read the entire book, depending on where you are in the process, but if you understand the basic steps and are systematic, you can be successful. Table 1 gives a overview of the process and indicates which chapters provide information about a particular topic.

Take an organized, objective approach to implementing your wastewater treatment system. Determine your wastewater management limitations and characterize your waste stream. Develop wastewater management alternatives and evaluate them. Select the best alternative and implement the wastewater treatment system, if you have to. For advice, help, and ideas, talk with others who have put in systems, work with vendors, read books and literature. Finally, meet with the agencies that regulate your facility before you invest a lot of time and money.

Table 1 How to Find Your Way Through This Book

Read:	To:
Chapter 1: How to Use This Book	Get a general overview of what information this book contains and how to approach it
Chapter 2: Characterizing Your Waste Stream	Determine the properties of your wastewater that you need to know to design a treatment system and to review sampling methods
Chapter 3: Determining Your Wastewater Management Limitations	Review the basis of discharge regulations and their relationship to other regulations
	Consider the budgetary, part quality, and work-flow requirements that will affect your decision
Chapter 4: Developing Wastewater Management Alternatives	Find an approach to identify and define options to bring your wastewater disposal into compliance with your discharge limitations
Chapter 5: Evaluating the Alternatives	Develop objective information about each alternative
	Conduct field and pilot tests to confirm claims made about an alternative
Chapter 6: Selecting the Best Alternative	Compare and rank alternatives to determine which one best fits into your shop's operations
Chapter 7: Implementing the Wastewater Treatment System	Prepare a plant layout, a process design report, plans and specifications, operations and maintenance manual, obtain permits, train shop personnel, and install the equipment
Chapter 8: Working with Suppliers	Increase your chances of a successful interaction with vendors and consultants and reduce your risk of making a bad purchase
Chapter 9: Annotated Bibliography	Find out about some wastewater pretreatment-related books that provide more information to help you design your system
Appendices	See examples of three treatment systems which include an explanation of their development and operation with installation and operating costs

Examples of three systems that were installed to treat relatively small discharges are given. They include treatment of oily and metal-bearing wastewater by evaporation or chemical precipitation. Both small and large facilities may be able to apply them directly to their situation. The small-scale examples are given because much of the available literature deals with the treatment of larger flows and large facilities. The processes, especially the chemical precipitation process, can readily be scaled up for larger flows.

BOOK OUTLINE

Chapter 2: Characterizing Your Wastewater

This chapter explains how to determine how the waste is generated, how much is generated and what it contains. The waste stream must be characterized to determine if it will meet regulatory requirements, and to design a process which can effectively reduce, recycle, or treat the waste if it does not. If the waste is not adequately characterized, the chosen method of waste minimization, treatment, or disposal may not work at all.

At a minimum the following must be known: (1) rate of waste generation and waste volumes, and (2) contaminants and their concentrations, including pH, metals, and organics content.

Chapter 3: Determining Your Wastewater Management Limitations

You must know your limitations. Before you decide how to proceed with your wastewater problem, make sure you understand its nature. Perhaps the most important restriction is compliance with wastewater discharge limitations.

Chapter 3 summarizes the potential regulations, process concerns, and economic considerations that define the boundaries of wastewater management decisions. The characteristics of the waste must be compared to the discharge limitations (if the waste is to be sewered), the air pollution control requirements (if the waste is emitted to the atmosphere), or the disposal criteria (if the waste is to be landfilled or incinerated). If the waste is to be recycled it must be acceptable to the recycling company.

The following items must be considered:

- Shop constraints
  - Your budget
  - Part quality requirements
  - Workflow requirements
- Regulations pertaining to
  - Sewer discharge

- Solid waste disposal
- Hazardous waste disposal
- Air pollution control
- Industrial hygiene and safety

Recycler's requirements

Treatment, storage and disposal facility (TSD) requirements

Chapter 4: Developing Wastewater Management Alternatives

Before building a treatment plant, you must be certain a treatment plant is the best alternative. Other alternatives should be identified. Can the waste stream be modified by a change in the industrial process? Would it be more cost effective to ship the waste offsite to a TSD for treatment?

If a wastewater treatment system is the most viable alternative, your next step is to identify the most applicable treatment technologies. Resources that can help you to determine what technologies are available are described. Many suppliers offer a variety of treatment equipment. Reference books, textbooks, and U.S. Environmental Protection Agency (EPA) manuals suggest an assortment of wastewater treatment methods.

The methodology used to develop wastewater management alternatives includes:

- Changing the process to reduce or eliminate the waste stream
- Segregating waste streams to reduce the volume of waste to handle and/or allow effective treatment
- Identifying applicable technology
  - Examine case studies
  - Contact suppliers
  - Contact shops that produce wastes similar to yours
  - Use the library to find applicable articles
  - Use trade associations
  - Contact regulatory agencies for information

It is important to get an overview of the available technology. Having a variety of alternatives to compare and contrast can help you to make an informed decision. Every technology has its drawbacks. You need to be able to choose the one that benefits you the most.

Chapter 5: Evaluating the Alternatives

This chapter explains how to go about evaluating the identified wastewater management alternatives. Before making your selection:

- Determine the reason for building the plant
- Examine alternatives to building the plant

Visit other shops that have systems similar to yours
Bench test selected technologies
If needed, pilot test the most promising processes
Consider
  Effectiveness and economics
  Capital and operating costs
  Space requirements
  Is the equipment a good fit for the technical level of the shop personnel?
  Reliability of the system
  Flexibility of the system
  Quality of technical support from vendor
  Disposition of waste treatment residual
Compare costs and benefits of in-house treatment vs. sending the waste to an off-site treatment company
Do not rely on hearsay; check it out yourself

An ill-conceived treatment system may not work, may not be cost effective, may be a nightmare to run, or may look so bad that it does not gain the confidence of a visitor. Following the outlined steps will reduce the risk associated with putting in a wastewater treatment system.

Chapter 6: Selecting the Best Alternative

This chapter examines the selection of an appropriate wastewater management alternative based on the particular needs of your shop. It explains how to do a cost assessment and how to prepare a system design. You will probably want to choose the most cost-effective alternative which has the highest probability of keeping you in compliance with the regulations.

Chapter 7: Implementing the Wastewater Treatment System

This chapter takes you from the concept to the hardware on the shop floor. To implement the system the following must be accomplished:

Prepare a plant layout
  Size the equipment according to waste volume information
  Provide spill-containment structures
  Comply with building and fire codes
  Ensure that the system is operator-friendly
Prepare a report that documents the system design
Prepare plans and specifications
Prepare an operations and maintenance manual
Obtain the necessary permits
Train shop personnel
Install the equipment
Start up and evaluate the system
Document system performance

The purpose of the design report is to document the work done to develop or redesign the waste treatment system. The report should be complete and detailed enough that a person familiar with waste treatment can assess whether an adequate job was done in selecting the waste treatment process chemistry and equipment. The report should contain, at minimum, a concise summary of the characteristics of the waste stream, a description of the procedure used to select the treatment process, and a description of the treatment process.

Plans and specifications for the waste treatment system should be prepared. The plans and specs are used to ensure that the finished plant will be built to minimum standards, to obtain supplier quotations, to allow the construction manager to determine if the plant is built correctly, and to satisfy a regulatory agency that the plant will be built as intended. The plans and specs should be complete enough that a supplier can install the waste treatment system as intended by the designer. The level of detail supplied in the plans and specs will depend on the specific project.

An operations and maintenance manual should be prepared before the treatment plant is started up to provide a guide for the plant operator. The purpose of the manual is to present technical guidance and regulatory requirements to the operator to enhance operation under both normal and emergency conditions.

Chapter 8: Working with Suppliers

You may rely on wastewater treatment equipment and chemical suppliers to help put your system together. You probably will be dealing with suppliers for replacement parts and operating material such as chemicals or filters. Even if you decide to contract out the entire operation, including the operation and maintenance of the system, you still should not take their proposals and statements on blind faith. The purpose of this chapter is to give an approach to working with suppliers, reducing the risks involved in buying equipment, and ensuring that the equipment works as intended when it is installed and running.

When dealing with suppliers keep the following in mind:

They have solutions looking for problems
You have a problem looking for a solution
Define your problem before you contact the suppliers
Use them as a resource; they should ask for pertinent information
It is your job to assess the technology; the suppliers' job is to bring it to your attention
Ask for supporting data and verify them
Visit shops similar to yours that are using the suppliers' products

Chapter 9: Annotated Bibliography

Helpful wastewater treatment books are described. Publisher information is provided to help you obtain a copy. The books are written from a variety of perspectives and should help you to learn about the various factors involved in treatment plant selection, design, and operation.

Appendices

This book presents methodology which will be useful for the development of a treatment process in a variety of circumstances. Several shop examples are provided (Appendices 1, 2, and 3) in order to show how wastewater discharge requirements, process requirements, shop expertise, and pollution prevention play into wastewater management. The examples include collection and treatment of hull wash wastewater at a boatyard, minimization and evaporation of wastewater at an automotive machine shop, and the treatment of oily wastewater collected by vactor trucks and produced by bus maintenance operations. The technology shown is applicable to the treatment of wastewater generated by a variety of businesses and utilizes simple equipment and processes. Different technologies and processes can be evaluated using the same principles, and the examples can be used to provide a basis of comparison to generate decision-making information.

2 CHARACTERIZING YOUR WASTEWATER

INTRODUCTION

This chapter explains how to develop an understanding of how wastewater is generated, how much is generated, and what it contains. Wastewater must be characterized to determine if it will meet regulatory requirements, or to collect the information needed to design a process which can effectively reduce, recycle, or treat the waste, if it does not. If the waste is not adequately characterized, the chosen method of waste minimization, treatment or disposal may not work at all. Table 2 presents the key points of this chapter.

Regulatory agencies require that you characterize and designate your waste. You must know whether it is hazardous waste and dispose of the waste properly. You must also know if it complies with local discharge limitations if you plan to sewer it. The regulatory requirements for designating your waste are based on federal regulations, but do vary from state to state and from one local jurisdiction to another. The information you need to characterize your waste for treatment on-site will help you to designate the waste. However, you need to know more about your wastewater than whether it is hazardous waste in order to treat it.

The wastewater characterization procedure explained in this and in the treatment system evaluation chapters is not intended to comply with regulatory requirements for hazardous waste designation. It is intended to help you to understand your wastewater well enough to successfully treat it. Refer to reference books that deal with hazardous waste designation for compliance with hazardous waste regulations. Contact your local pretreatment authority for guidance on local discharge limitations. Use this book to develop an understanding of how to characterize your wastewater for treatment and how to present a well-documented report to your local pretreatment program staff to obtain authorization to discharge your wastewater to their sewage system.

One cause of treatment system failures is inadequate characterization of the wastewater. You must understand the nature of your wastewater to treat it effectively. The important parameters include the volume of the wastewater, the pattern of generation of wastewater (all at once, evenly through a shift, or

Table 2 Key Points of Wastewater Characterization

Understand how the wastewater is generated
Know the processes in your shop that contribute regulated contaminants
Know what is in the products you use
Know what is being removed from the material being processed
Measure how much wastewater is produced
Determine what regulated contaminants are in the wastewater
Understand how much the volume and contaminants vary over time
Find out what others know about wastewater that is similar to yours
Sample your wastewater to verify what it contains
Prepare a sampling plan; think out what you are doing
Ask your pretreatment authority about their data reporting requirements

in fits and starts), and the chemistry of the wastewater (what it contains and how it reacts).

You can approach wastewater characterization logically. You need to get a feel for your wastewater. If you understand how wastewater is generated, you are more likely to be able to treat it effectively than if you do not pay attention to it. When you experience the inevitable treatment plant failures and upsets you will have a better chance of determining what went wrong. For example, the failure may be a result of an improper treatment chemical dosage or it may be a result of pouring concentrated process solution into the treatment system. If you have a sense of what your wastewater looks like then the cause of variations in the treatment process will be more evident.

Table 3 indicates the wastewater parameters you need to know to design and operate a wastewater treatment system. The terms are defined and explained in this chapter. The methods used to determine the parameters are also given and explained.

THE SOURCE — WHERE IS THE WASTEWATER GENERATED?

You are familiar with the work that you do in your shop. You know the steps to produce your product. You follow a logical sequence. You can apply the same approach to looking at the way wastewater is generated in your shop. Look at your wastewater the same way that you look at your product. Get familiar with it.

You need to develop a foundation to work from and an organized approach. An overview of your shop's operation is needed. You do not want to get mired in details before you have determined what is important to worry about.

Inventory the processes your shop performs and make a list of them. This list is the starting point for characterizing your waste stream. Wastes are generated as a result of processes. Take a process view of your operation. A process is, "a particular method of doing something, generally involving a number of steps or operations" (*Webster's Unabridged Dictionary,* 2nd ed., 1983). Walk through your shop, think about the process steps that are utilized.

Table 3 Wastewater Parameters You Need to Know

The source — where the wastewater is generated
The volume — how much is produced
The generation pattern — when it is discharged
The contaminants — what it contains, including:

- pH
- Flash point
- Temperature
- Heavy metal concentrations
- Regulated organic compounds present and their concentrations
- FOG concentration
- BOD, if required
- Other contaminants of concern to your local pretreatment authority
- The variability, how much the above parameters change

Table 4 Example Shop Process Lists

Shop	Processes
Radiator	Draining the cooling system
	Flushing the cooling system
	Removing the radiator
	Cleaning the radiator
	Repairing the radiator
	Testing the repair
	Installing the radiator
	Cleaning the floor
Machine	Initial cleaning of an engine block
	Machining the block
	Cleaning the block after machining
	Protecting the block from rusting
	Cleaning the floor

How does the work flow through your shop? Table 4 is an example of lists which may be applicable to radiator shops and machine shops.

The processes and their relationships may be easier to envision if you produce a flow chart. The flow chart is an outline of the operations used in your shop. It can be used in association with a layout drawing of your shop that shows where each of the process steps occurs. An example shop layout and flow chart are shown in Figures 1 and 2.

You should produce a similar process list and layout for your shop. Include the major process steps. The next step is to consider each process and determine if wastewater is produced as a result of it.

Rinsewater will probably be the largest single volume of industrial waste produced by your shop if you are using aqueous processes. Note where rinsewater is produced. For example, include aqueous process steps such as caustic hot tank cleaning, acid cleaning, plating, radiator flushing, equipment cleaning, and floor cleaning. Organize the rinsewater-generating steps using the process step list.

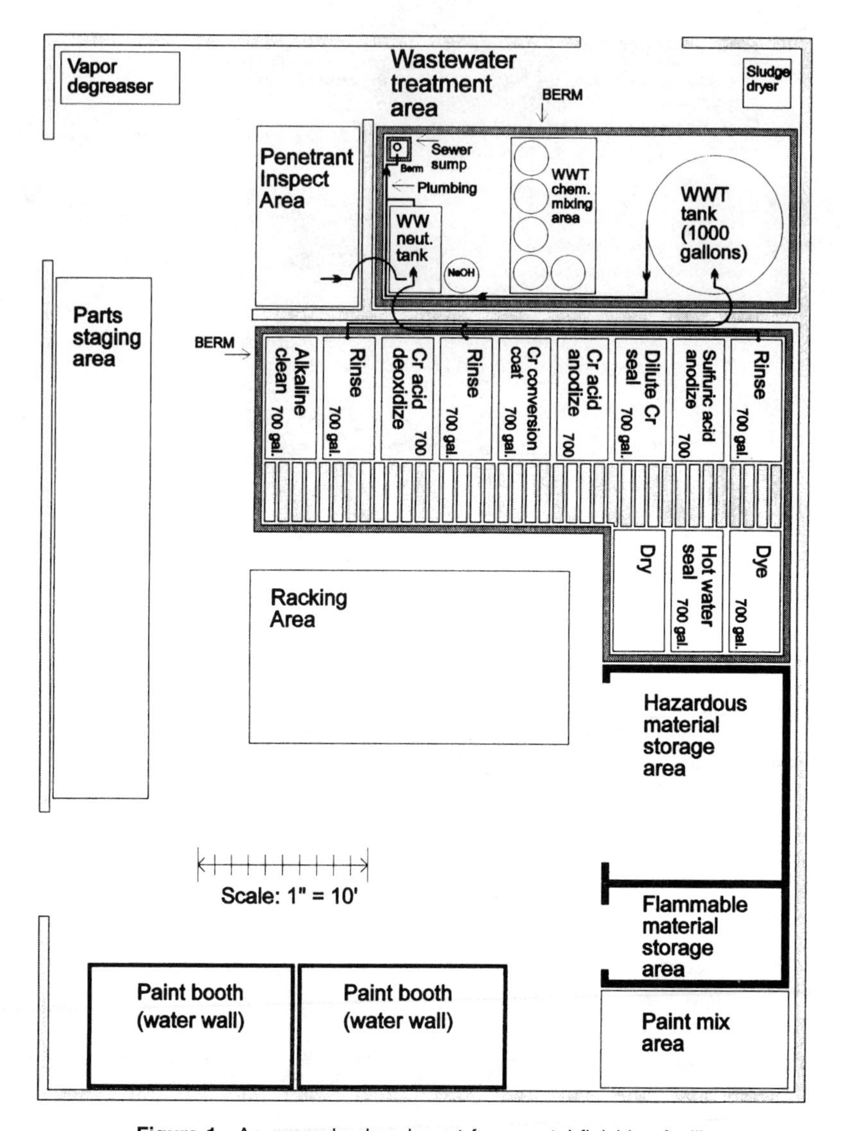

Figure 1 An example shop layout for a metal finishing facility.

You should also list all of the aqueous process solutions, such as machine coolant, found in your shop whether or not they are directly associated with a rinsing step. The aqueous process solutions get onto the parts being processed and are removed in subsequent processes. The machine coolant will be removed in an aqueous or solvent cleaning process and will end up as a contaminant in the waste produced by the process. In addition, you may want to consider reclaiming and reusing the process solutions or treating them in-house and discharging them rather than sending them off-site for disposal.

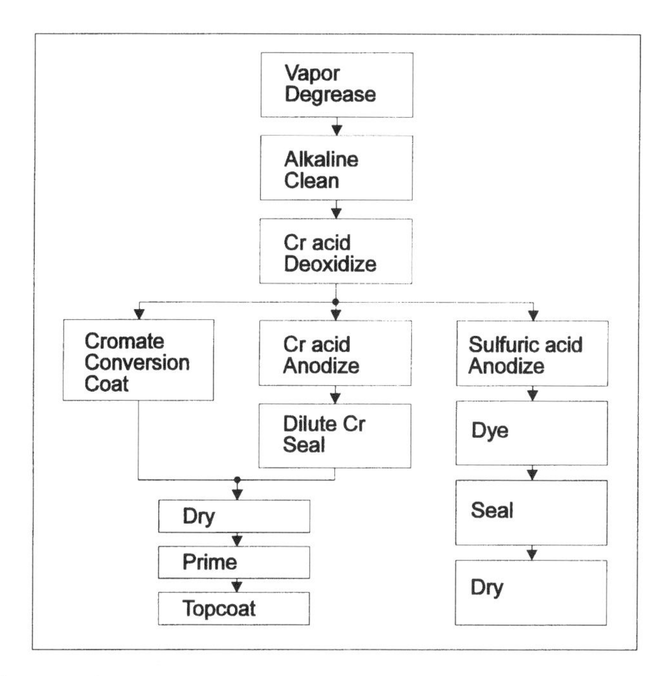

Figure 2 An example process flow diagram for a metal finishing facility.

After you have determined where you produce wastewater, you need to determine how much is generated and the pattern of generation.

THE VOLUME — HOW MUCH WASTEWATER IS PRODUCED?

The size of a wastewater treatment system is dependent on how much wastewater must be treated. You should not install a system without knowing how much water you will have to treat. You must know the average and peak daily wastewater flow rate. Depending on how you design your treatment system, you may need to know how the flow rate varies over the course of a day. Measuring the flow rate depends on how the wastewater is produced and managed in your shop. Water flow rates can be measured using a flow meter, the bucket method, or the tank method.

Your water bill can give you an idea of the average amount of water you use for all purposes in your shop because your water use is metered. However, the usage is given as a monthly total and daily average based on the monthly total and includes water used in your restrooms, for landscape watering, and other nonindustrial uses. Therefore, the water bill will only give you an indication of the average upper limit of your monthly water usage. It does not show the daily ups and downs of water usage. To get daily usage numbers you have to read the meter yourself and keep a record. Also, your water meter may service other businesses at your location. If so, then you can not differentiate the amount you use in your shop.

If you are on your own water meter and use most of the water in your industrial processes, the water bill may provide useful information. For example, a plating shop which has 20 employees and uses about 10,000 gal/day of rinsewater may discharge 700 gal/day of water from the restrooms (about 35 gal per employee). In this case, because most of the water used is used in the industrial processes, the water bill will reflect the average industrial wastewater volume. Daily reading of the water meter will closely reflect the amount of water used that day in the plating line and associated processes.

On the other hand, the water bill can give misleading information. For example, a water bill from a machine shop with about 30 employees indicated that some 2000 gal/day of water was used by the facility. However, this shop only uses 200 to 300 gal/day of water in its industrial cleaning operation, based on flow measurements at the process area. In this case, the industrial process contact water was a small fraction of the total water used. The additional water was used in the restrooms (about 1000 gal) and to cool the compressors. If the size of the treatment system had been based on the water bill volume, the system would be much larger than needed.

You can use a dedicated water flow meter to measure how much water flows into a process. If most of the water is discharged and neither evaporated nor incorporated in your products, measuring how much water is going into the process will give an adequate measure of how much water is discharged. If you use a hose for rinsing parts, it is easy to install a flow meter in the line to get a direct reading for that specific operation. If you are using a flow-through rinse tank, you can install a flow meter on the tank water inlet to determine how much water is used in the tank. If you have several process areas or tanks you could purchase one meter and rotate it between locations to develop water use information for each location.

City water meters are readily available and are in general the least expensive type of totalizing flow meter. The meter you choose must totalize flow. It is desirable for the meter to show the instantaneous flow rate, especially if it is installed on a rinse tank. The water flow for the process can then be set easily using the meter. This provides a means of regulating water flow so that your treatment system is not overwhelmed by excessive flow rates.

Figure 3 shows a flow meter installed in a sink used to clean printing screens. The flow meter shown has a working range of 0.2 to 20 gal/min and shows the instantaneous flow rate and also totals the flow. It cost approximately $300. You may be able to find one that costs about $150. (The fittings to install the flow meter were purchased at a local plumbing store for about $50.)

The bucket method can be used in a situation in which the rinsewater flow is the same each day and constant throughout the production day. To use this method, place a bucket of known volume at the rinsewater outlet to collect all of the water coming from the process and record the time needed to fill the bucket. Use a container that takes at least 1 minute to fill to get an accurate reading. Calculate the

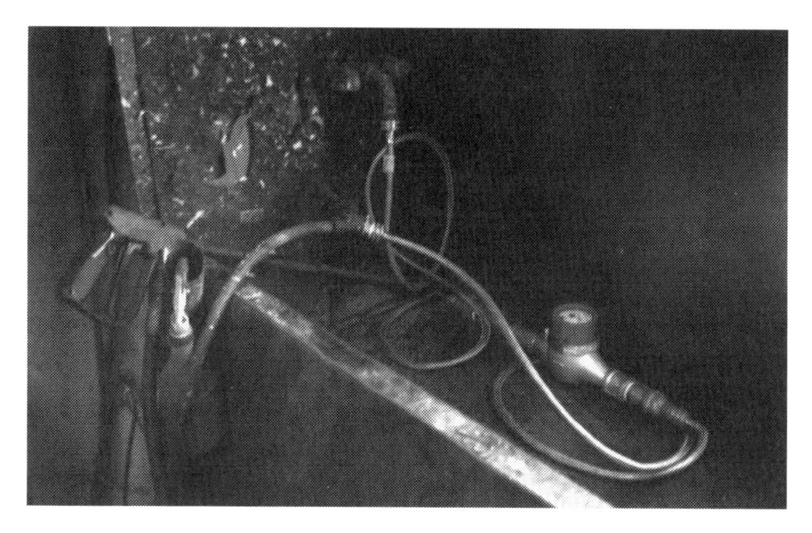

Figure 3 A flow meter installed at a sink used to clean printing screens.

flow rate in gallons per minute by dividing the volume of the bucket in gallons by the time it took to fill the bucket in minutes. Repeat the measurement several times on different days to ensure that the reading is reproducible and accurate.

If you have a rinse tank that is hard plumbed to the drain, you can use the bucket method without disconnecting your plumbing. Take a 5-gal bucket and submerge it in the tank. Wait until the tank is overflowing normally again and lift the full bucket from the tank. Now record the time it takes for the tank to fill to overflowing again. The water running into the tank during that time equals the volume of the bucket. Calculate the flow rate the same way as if you measured the time it took to fill the bucket, by dividing the volume by the time.

The tank method is essentially a big bucket method. Collect all of your wastewater in a tank and measure the volume collected over time. Use a sump pump to pump the wastewater into the tank. The flow rate is equal to the volume collected divided by the time it took to collect it. The size of tank you will need is, of course, dependent on the amount of wastewater you generate, so you will have to estimate the amount of wastewater you produce before you obtain the tank. The tank can range in size from a 30-gal plastic trash can to a 55-gal drum to a 1000 gal or larger plastic or metal tank.

If you have a large sump you could use it as a measuring tank. First empty the sump, disposing of the contents appropriately, measure the sump size, and calculate the volume:

$$\text{Volume in cubic feet} = \text{length in feet} \times \text{width in feet} \times \text{depth in feet} \qquad (1)$$

Then volume in gallons:

$$\text{Volume in gallons} = \text{volume in cubic feet} \times 7.48 \text{ gal/1 ft}^3 \qquad (1a)$$

Then measure how long it takes to fill the sump and calculate the flow rate. If the sump takes longer than 1 day to fill, then measure the depth of the water at intervals to get a daily flow rate.

THE GENERATION PATTERN — WHEN IS IT DISCHARGED?

In a typical shop wastewater is produced sporadically. It is not produced at an even rate. You should determine the wastewater generation pattern in your shop. Do you produce about the same amount of wastewater each day or do you generate 1000 gal in one shot every other week?

How often you read the meter depends on how it varies and how you intend to treat the wastewater. If you are going to batch treat the wastewater then you must know how big a batch you will be treating and how long it will take to collect the batch. How much the flow rate varies while the batch tank is filling does not matter. What matters is how often the tank fills up. If you plan to install a continuous treatment system then flow variations are more of a concern. The continuous system must be able to keep up with the flow of wastewater under the peak discharge flow conditions. For example, you may have a flow rate of 200 gal/day and need a 200-gal batch treatment tank. It is sufficient in this case to read your water meter once a day. However, you need more information if you use a continuous treatment system. Is that 200 gal generated evenly over the production day or is it produced in 20 10-gal spurts? Batch and continuous treatment and their differences are discussed in a later chapter.

How long a period — days, weeks, months — that you measure the flow rate depends on how much your production fluctuates. You need to judge whether production is up, down, or normal when you are measuring your wastewater volume. You have a sense of how typical the day is when you are measuring the water flow rates. This is where your judgment comes in. If you are installing your treatment system during your normally slow time of the year you must account for that factor. Obviously, if you measure the wastewater volume during a slow period and size your treatment system for that flow, it will be too small to keep up with normal production flows. The best solution is to measure the wastewater generated at various levels of production and size the system to handle full production.

The intent of actually measuring your wastewater volume is to reduce the risk of installing a treatment system that is too small or that is much bigger than it needs to be. Uncertainty is involved in the flow measurement. The uncertainty includes the variation in production levels and processes used and the consequent variation in water usage. Ideally, you want flow information for various levels of production. Practically, you may not be able to measure flow rates for an extended period of time. Thus, you may be able to assume that water usage follows production. If production doubles then water usage doubles. This assumption is good if water usage depends on the number of items processed. It does not hold if you have a rinse tank that you keep running at the same rate regardless of what you put through it. In that case water flow does not vary with production.

If you intend to implement water conservation methods, such as a rinsewater conductivity controller, which adds water to a rinse tank in response to the load going through the tank, you should take into account the anticipated wastewater reduction when sizing your treatment system. The development of water conservation methods is discussed in a later chapter.

In summary, measure the wastewater flow rate using the flow meter, bucket, or tank method. Take readings often enough to see how much the flow rate varies during a production day and from day to day. Continue to measure for as long as possible to get as much information as you can. Account for variations in production by estimating the flow at full production. Factor in anticipated reductions in flow rate due to water conservation methods.

THE CONTAMINANTS — WHAT DOES THE WASTEWATER CONTAIN?

In order to determine what to do with your wastewater you have to know what is in it. Discharge regulations require that you know what you are putting down the sewer. Your choice of pretreatment technology depends on what you are removing from the wastewater. The parameters of concern to industrial pretreatment program include those given in Table 5. Discharge limitations for those and other parameters depend on the conditions particular to your locality. Ask your local authorities about their requirements. A regulatory overview is provided in Chapter 3. A description of the parameters is presented here.

Table 5 Wastewater Properties Regulated by Industrial Waste Pretreatment Authorities

Ammonia (as nitrogen) concentration
BOD
Chemical oxygen demand (COD)
FOG concentration: polar and nonpolar fractions
Flash point
Fluoride
Nitrate and organic nitrogen (as nitrogen)
pH
Regulated organic compounds and their concentrations including:
Pesticide active ingredients (PAI)
Volatile and semivolatile organics
Total phenol
Total toxic organics (TTO)
Temperature
Total cyanide concentration
Total metal concentrations, including antimony, arsenic, beryllium, cadmium, chromium, hexavalent chromium, copper, lead, manganese, mercury, nickel, selenium, silver, thallium, and zinc
Total organic carbon (TOC)
Total phosphorus (as P)
Total suspended solids (TSS)

Two main considerations determine what is in your wastewater: What is there and how much does it vary? You have to select the appropriate test method to determine what is there. Sample collection and testing methodology depends on what you are looking for. The nature of your shop's processes affect how much the wastewater composition varies.

Determining what is in your wastewater and how much the concentrations of the components vary is a challenge. What and how much is in the wastewater can vary widely depending on the process. You should approach the problem from two directions. Look at what may go into the wastewater based on what products you use and what kinds of material you process. You may not always know what is in the products you use or the materials you process. Therefore, you should sample the wastewater to confirm what is there and determine the concentration of the components present.

Wastewater Parameters of Concern

The following is a brief summary of important wastewater parameters. Additional parameters may be of concern in your district. Consult the pertinent references in Chapter 9 for more details.

pH

pH is defined as the negative log of the hydrogen ion concentration. The wastewater pH can be measured with a pH meter or pH paper. The amount of acid or base affects the pH of the water. A neutral pH, that of pure water, is 7.0. A pH below 7 is on the acid side and a pH above 7 is on the basic side. Table 6 gives the pH of some common solutions. The pH of the listed foods may vary over a wide range.

A pH of below 6 or above 9 can upset a municipal treatment plant or a septic tank. Wastewater with a pH below 5 will corrode a concrete sewer, leading to failure of the sewer and an expensive repair. Municipal sewage

Table 6 pH of Some Common Solutions

Solution	pH
5% Nitric acid (HNO_3)	0
Lemon juice	2.2
Vinegar (5% acetic acid)	2.5
Soda pop	2.5
Orange juice	3.2
Beer	4.5
Milk	6.2
Pure water	7.0
Sodium bicarbonate (baking soda) solution	8.3
1% Ammonia	10.2
Lime, saturated solution	12.5

Table 7 Flash Point of Some Common Liquids

Liquid	Flash Point
Acetone	15°F
d-limonene (>1% aqueous solution)	~130°F
MEK	24°F
190-Proof ethanol (95%)	55°F
Isopropyl alcohol	59°F
Trichloroethylene	90°F
Gasoline	–50°F
Kerosene	100–150°F
Diesel oil	110–190°F
1,1,1-Trichloroethane	None
Water	None

typically has a pH of between 6 and 7. The pH of the sewage is usually easier to drop with additions of acid than it is to raise with the addition of base. It is buffered on the basic side. Therefore, acidic wastewater is more detrimental to a municipal treatment plant sewage than is basic wastewater. Pretreatment authorities strictly regulate pH.

Flash Point

The flash point of a liquid is the temperature at which its vapor will ignite when exposed to an ignition source. ASTM methods D93 or D3278 are commonly used to determine flash point of water. The lower the flash point the more flammable the solution. A flash point instrument is required to perform the analysis. The flash point of some common liquids is shown in Table 7. The flash point test (ASTM D93 or D3278) was developed for use on fuels and is adversely affected by the presence of water, but applies to wastewater nevertheless.

Federal regulations prohibit the discharge of any wastewater with a flash point of <140°F. The presence of flammable liquids and vapors in a sewer can result in an explosion. You may remember the incident in Mexico in 1993 in which a gasoline leak into a sewer resulted in widespread damage to sewers, streets, and buildings when the vapor ignited.

Temperature

Heat is added to incoming water for industrial processing, sanitation, and cooking. Therefore, the temperature of sewage is typically higher than tap water. The temperature of wastewater is easily measured with a thermometer.

Hot water with a temperature of >150°F can injure sewer workers. The bacteria at the municipal treatment plant prefer a temperature of between 60 and 100°F. If the temperature gets higher or lower then the process functions less efficiently. Therefore, your wastewater discharge temperature is regulated.

Heavy Metal Concentrations

Metals cannot be degraded in a treatment plant. They either end up in the treatment plant sludge (also known as biosolids) or pass through the plant and leave with the effluent. Certain metals are regulated because they are toxic to humans and animals and can adversely affect the municipal treatment plant operation. Regulated metals include arsenic, barium, cadmium, chromium, copper, lead, mercury, nickel, selenium, silver, and zinc.

Municipal treatment plants have strict effluent metal concentration limitations. The sludge produced by the treatment plants must be disposed of, but disposal options are limited. The most beneficial use of the sludge is land application on forests, farms, and landscapes. The land application of sludge is strictly regulated by the EPA and metals content of the sludge is closely monitored. Large industrial dischargers have been subject to wastewater discharge permit limits for years, resulting in a large decline in the metals concentration.

It appears that most of the metals currently received by large municipal treatment plants are from residences and small businesses, which are not required to have industrial wastewater discharge permits. Pretreatment authorities are paying more attention to metal discharges from small businesses and residences as municipal treatment plant effluent limitations decrease.

Metals are typically analyzed using an atomic absorption spectrophotometer (AAS) or an ion-coupled plasma spectrophotometer (ICP). Metals are identified using the characteristic light waves emitted when the metal atoms are excited by the instrument. The wastewater metal concentrations are determined by comparing the instrument response to the wastewater to its response to standard solutions with known metal concentrations. Refer to the appropriate reference in Chapter 9 for analytical method details.

Organic Compounds Present and Their Concentrations

In the context of wastewater treatment, organic compounds comprise a broad range of substances, including common industrial solvents, disinfectants and other pesticides, gasoline, resins, polychlorinated biphenyls (PCBs), dioxin, chemical reagents, and chemical process intermediates. The environmental risk and fate of the compounds varies and depends on their properties. Each class of organics must be looked at individually because of the wide variety of compounds and their different properties. Within each class of compounds, each organic may need to be evaluated.

In addition to the properties of the compounds, local conditions must be taken into account, including the wastewater volume and concentration of the compound discharged, the size of the municipal treatment plant, and the treatment plant discharge and sludge quality requirements. Contact your local pretreatment authority to determine if your organic contaminated discharge is acceptable.

In a municipal wastewater collection and treatment system the organic compounds may volatilize into the air, be biologically degraded, pass through the treatment plant into the effluent, or end up in the sludge. Several examples of the primary fate of organics follow. Compounds such as PCBs and dioxin are not volatile, degrade very slowly, and are not soluble in water. Therefore, they tend to end up in the treatment plant sludge. Compounds such as acetone and methyl ethyl ketone (MEK) are volatile but very soluble in water and are amenable to biological treatment. Therefore, they are degraded in the treatment plant. Compounds such as toluene and xylene are volatile, not water soluble, and are amenable to biological treatment. They tend to air strip from the wastewater as it flows through the sewers. Toluene and xylenes that make it to the treatment plant are removed biologically or by air stripping. Compounds such as 1,1,1-trichloroethane and trichloroethylene are volatile, not water soluble, and not amenable to biological treatment. They are air stripped in the sewer or at the treatment plant.

Volatile organics that air strip in the sewer or at the municipal treatment plant may present a health and safety risk to workers and the public. The presence of PCBs or dioxin in the treatment plant sludge can restrict the disposal options and increase the cost of sludge management. Organics that pass through the treatment plant can contaminate receiving waters, resulting in environmental damage and drinking water contamination. Therefore, the discharge of organic compounds is regulated.

The analytical method used to quantify an organic depends on the class of the compound. Volatile organics are analyzed by purging them from the wastewater with nitrogen gas, collecting them in an adsorption column, and heating the column to desorb the organics into a gas chromatograph (GC), which separates the organics for identification and quantification. The type of detector used with the GC depends on the compound and the level of confidence needed. The most rigorous method uses a mass spectrometer (MS) as a detector. Hence, you may hear the method referred to as the "purge and trap" method using a GC or a GC/MS. Organics that are not volatile are extracted and concentrated using various methods and then run through a GC or GC/MS. Use the appropriate reference in Chapter 9 for analytical method details. Your pretreatment authority and laboratory can help you select the best method for analysis of your wastewater.

Fat, Oil, and Grease (FOG) Concentration

Much of the FOG released from industries and residences is emulsified (made soluble in water) by detergents and soaps used for removing it from whatever was cleaned. Some of the FOG is released as flowable oil such as when used motor oil or cooking oil is poured down the drain, and some is released from food preparation activities as solidified fat globules.

A portion of the FOG tends to float on the surface, forming grease balls which interfere with treatment plant operations. It also clogs pipes, making periodic disassembly and cleaning necessary. The fats and grease also adhere to and are removed with the sludge.

Fats, oils, and greases are classified into two major categories: polar and nonpolar FOG. Polar FOG tends to be of vegetable or animal origin. Milkfat, vegetable oil, lard, and peanut oil are examples of polar FOG. Nonpolar FOG tends to be petroleum based. Motor oil, petroleum jelly, diesel oil, and axle grease are examples of nonpolar FOG. Nonpolar FOG is not readily broken down in the municipal treatment plant and is strictly regulated. Polar FOG is more amenable to biological treatment and therefore tends to be less strictly regulated than nonpolar FOG.

The analytical method for FOG entails extracting the FOG from the wastewater with a solvent. Trichlorotrifluoroethane (a Freon) is the standard solvent, but is in the process of being replaced by another solvent. The FOG in the extract is determined in one of two ways, gravimetrically or spectrophotometrically. The method of choice depends on the nature of your wastewater and the requirements of your local pretreatment authority.

Biochemical Oxygen Demand (BOD)

The BOD is a measure of the amount of oxygen used by treatment plant bacteria to oxidize the organic content of the wastewater. In this case organics include all carbon-containing compounds in the wastewater, e.g., sugars, carbohydrates, and manufactured organic chemicals. The BOD measurements are used to design and control municipal wastewater treatment plants.

A biological treatment plant is designed to remove BOD from wastewater. However, a biological treatment plant can be overwhelmed and caused to fail if it receives waste with a high BOD. Biochemical oxygen demand is measured by incubating wastewater with a bacterial seed and measuring how much oxygen is used over a 5 or 7 day period.

WHAT IS IN THE PRODUCTS YOU USE?

For each process that generates wastewater, list what products and process chemicals are used. Look at the Material Safety Data Sheets (MSDSs) that the chemical supplier sends to you along with the process chemicals. The MSDSs are required to list components that are regulated under various laws. Keep in mind that MSDSs are regulatory documents prepared by manufacturers and are not typically independently reviewed. Use them with caution.

The best MSDSs list all of the product components and their concentrations, include accurate disposal information, and give product toxicity information. However, MSDSs are not of consistent quality. Some give no product component information, stating that it is a trade secret. In this situation you are in the position of having to trust the manufacturer's claims that no regulated materials are in the product. This may be the case, but it would be better if you knew for certain what was in the product. Where possible, buy your products from a manufacturer who will tell you what is in the products that you use.

WHAT IS REMOVED FROM THE MATERIAL BEING PROCESSED?

List what kinds of contaminants may be removed from the items being processed. It is not enough to know what is in the products that you use. The products may not need treatment before they are used, but require treatment after they have picked up the soil removed from the parts. For example, an aqueous cleaner may contain detergents that are acceptable for discharge without treatment, but if an oily part is cleaned with the detergent, the solution is then contaminated with oil and may exceed discharge limitations. Or rinsewater from a fresh caustic hot tank or jet washer solution may be acceptable for discharge to a municipal treatment plant or a publicly owned treatment works (POTW). However, in use in a radiator or machine shop, for example, the solution will pick up lead, copper, and zinc. These metals, then, will need to be removed from the rinsewater before it can be discharged. In an automotive shop the same hot tank solution will pick up lead and oil from the parts. Again, the wastewater produced will most likely require treatment prior to discharge to the sanitary sewer. Unused photographic fixer may be acceptable for discharge, but when it is used it removes silver from the processed film, and its characteristics change, restricting its discharge.

Also, note that wastes from processes which do not directly produce wastewater can get into the wastewater. For example, someone cleaning a part with solvent in the steam cleaning area may let the solvent run into the steam cleaning wastewater sump. If so, the wastewater will contain solvent which may need to be removed before it is discharged. It is usually more cost effective to keep the solvent out of the wastewater in the first place rather than remove it after it was needlessly put in. Understand the processes used in your shop and how they overlap. Look for cross-contamination and ways to prevent it.

Wastewater contaminants will vary between processes and shops. This book does not attempt to list all of the possible contaminants. You need to take a look at your processes with respect to discharge limitations and use your knowledge of what you do to determine what potential contaminants are of concern.

WHAT DO OTHERS KNOW ABOUT YOUR WASTEWATER?

Much information about wastewater is available. Before you go off on your own, you should contact other shops doing similar work, product suppliers and manufacturers, trade associations, and government agencies, such as your local pretreatment authority or hazardous waste program. They may have information which will help you to characterize your wastewater and get you started in the right direction.

Most shop owners are willing to share what they have learned about their wastewater. If they do similar work, what they know will be applicable to your situation. They can tell you how they determined what is in their wastewater and what is likely to be in yours. They can also tell you what laboratories and test methods they used to analyze their wastewater.

Product suppliers and manufacturers may have information regarding how their products are used and what contaminants are likely to be picked up. They may be able to provide characterization data beyond what is offered in the MSDSs, such as what a spent process solution looks like. The suppliers also may get you in contact with other shops using their products (see the chapter on working with vendors for guidance).

Your local pretreatment authority and hazardous waste program representatives are dedicated to understanding the nature of industrial wastewater. They have information about the characteristics of a wide variety of wastewaters and can help you determine what may be in your wastewater and can suggest appropriate test methods. They also are aware of shops similar to yours which have already characterized their wastewater. Some local hazardous waste programs sponsor annual "waste information" trade shows which include seminars on waste management. In addition, they may be able to identify trade associations which can help. Call your tap water supplier to find out what pretreatment jurisdiction you are in and who to call for industrial waste information. Once you have identified the pretreatment authority, they can suggest who to contact in your state or local hazardous waste program.

Your trade association may be a good resource. Many associations are active in environmental affairs. They can provide a focal point for information exchange. The local chapter members may have informative monthly meetings where you can meet other operators and shop managers and discuss waste management issues. Some of them make waste management a seminar topic at their national meetings.

In some cases wastewater properties and disposal options can be determined from existing information. Let us look at the silver-bearing fixer generated in the development of photographic film. Spent fixer is well characterized by manufacturers and fixer disposal options are available. Silver can be removed from the fixer in your shop, or, if you generate small amounts (a few gallons per week), another business that has silver recovery equipment may accept your fixer for treatment. In other words, when the process is well characterized and consistently maintained, you may be able to get all of the information you need to decide how to dispose of the wastewater. Therefore, try to find out what is already known before you decide how to characterize your wastewater.

SAMPLING YOUR WASTEWATER

Because of the difficulty of predicting what is in your wastewater, relying on product component knowledge to characterize your wastewater may be

Table 8 What a Sampling Plan Includes

What a Sampling Plan Includes
What waste streams will be sampled?
Who will take the samples?
What safety precautions will the sampler take?
How many samples will be taken?
When will the samples be taken?
How will the samples be obtained?
How will the samples be handled?
What methods will be used to analyze the samples?
What laboratories will do the analysis?
How will the data will be reported?

insufficient. In addition, you may have to sample your wastewater. Your local pretreatment authority can give you advice on what information they need to determine whether they can accept your wastewater.

Before you start sampling you should develop a sampling plan. The purpose of the sampling plan is to ensure that you effectively and efficiently get the data you need and can document that the data are representative. The elements of sampling plan are shown in Table 8.

The following text breaks down each element listed above to give some insight into how to develop your own sampling plan. The sampling plan has to be detailed enough to provide adequate direction to the person taking the samples. It should be documented well enough to be included in a report to the pretreatment authority. Ask them to review your sampling plan before you use it. They may make suggestions which will help you to satisfy their requirements. It is important for you to take some time to think out the plan. If you do not do a good job sampling your wastewater, you may end up installing a treatment system that is not needed, is bigger or smaller than it needs to be, or does not work at all. Spend some time up front at this stage in the developmental process to build a good basis from which to work.

What Waste Streams Will Be Sampled?

You have already determined which of your processes generates wastewater. You may have documented that some of the wastewater is acceptable for discharge without treatment and therefore does not need to be sampled. You should sample all the wastewater which you know or suspect contains regulated contaminants, including rinsewater and any process tank solutions, and that you plan to discharge to the sanitary sewer.

Who Will Take the Samples?

Decide who is responsible for taking the samples. In a small shop it is best to assign the responsibility for the entire waste treatment system project to one person. A larger company may assign a pollution manager to oversee a team of personnel who each conduct different aspects of the sampling program. The more people who are involved, the harder it is to coordinate the project and communicate task goals. If you are going to delegate someone other than the person responsible for developing the treatment plant, be prepared to explain in detail exactly what the sampling is for. Whoever takes the samples must understand how critical it is to get representative samples and how best to obtain one. If you are not sure how to take the samples, do not assume someone else knows how. It is not a good idea to trust to luck when sampling.

What Safety Precautions Will the Sampler Take?

Examine the sampling sites to determine what safety precautions are needed. Document the precautions. If a possibility exists that corrosive or toxic substances are present in the wastewater, waterproof chemical-resistant gloves and splash goggles should be worn. Look for physical hazards also, such as mixers, wires, and confined spaces. Make sure the sampler is aware of and protected against potential hazards.

How Many Samples Will Be Taken and When Will the Samples Be Taken?

This is one of the most difficult aspects of characterizing a wastewater. Ideally you want as many samples as you can get. However, sample analysis can be expensive and obtaining samples takes time. You can only get a limited number of samples.

Start by getting one set of samples. Collect them, have them analyzed, and review the data. Make sure that the samples are collected and handled properly and that you understand the results. This should serve to identify major problems in the sampling and analysis procedures, although you still may not know if you took the sample at the best time or are looking for the right compounds.

Compare the sample results to what you know about the chemicals you use, the contaminants that you remove from your work, and analytical results from similar processes in other shops. Then take another set of samples and compare them to the first set.

If your shop processes are fairly consistent, then the composition of the two samples should be similar. For example, two consecutive hot tank rinsewater samples with FOG concentrations of 700 and 1400 milligrams per liter (mg/L) are similar. If you have consistent processes which use the same chemicals each time, then several sample sets will give you an adequate idea of what the wastewater may contain.

Many businesses have a process tank such as a hot dip tank or a radiator test tank which is used for an extended period before it is changed out. In use the tank builds up contaminants, and consequently the rinsewater contamination from the parts processed in the tank increases. If you happen to collect your samples soon after you changed out the process tank, the samples will not be representative of the rinsewater generated when the tank is close to being changed out. In this case you must wait until the process tank is contaminated before you sample. It is best to sample the rinsewater when the process tank is about to be changed out. Your treatment system must be able to handle the rinsewater generated by a close-to-spent tank as well as by a fresh tank.

If the sample results vary widely, for instance, two consecutive samples of hot tank rinsewater with 0.5 and 50 mg/L of lead, then you better review your sampling technique. Determine if the variability is a result of the sampling or the processing.

Some wastewater is inherently variable. It can vary because the mix of chemicals and products varies throughout the day, week, or year. For example, during one shift, a truck-washing facility may clean out a truck that contained molasses, then one that contained vegetable oil, and then one that contained detergent. Or it can vary because the process removes varying amounts of contaminants from the work. For example, dentists remove mercury amalgam fillings from teeth. The amount of mercury removed depends of the size of the filling. The mercury found in dental wastewater from a single chair was found to vary between 50 and 800 mg/day, which is >10 times. Another cause of variation is that a certain job is done intermittently. For example, you may wash down your floor once a week, or a printer may wash printing screens daily and only reclaim screens once a week.

In these situations you must sample the wastewater enough to determine the range of variation. The truck-washing facility must sample rinsewater from trucks with each type of content. The dentist must sample daily for several weeks. You must sample your floor washwater when it is generated. The printer must sample the wastewater during routine days and on the days that screens are reclaimed. The number of samples needed depends on the situation. The main point is that you sample enough to have a reasonable certainty about the composition of your wastewater.

How Will the Samples Be Obtained?

Samples can be obtained in several ways, each having its merits and deficiencies. You can place a pan or bucket under the part(s) being rinsed and pour the collected water into a sample bottle. You can use an automatic sampler to take samples at predetermined time intervals over an operating day. You can manually take grab samples at predetermined intervals. You can collect all of the wastewater generated over the period of a day or longer and sample the mix. An example of how to apply each method follows.

Use a Pan or Bucket

This method is useful where the rinsewater runs onto the floor and into a dirty sump, or directly into a contaminated sump. In this situation a sample from the sump may not be representative. For instance, during hot tank cleaning a machine shop may generate oily rinsewater that contains metal-contaminated solids. In the sump, oil will float and the solids will sink. Solids will build up with time, and, depending on the design of the sump, oil may also accumulate. When you sample from the sump, you may collect a sample from near the top of the sump and think that the wastewater contains more oil than it does when it is produced because you sampled an oil-rich fraction. Or you may pull a sample from near the bottom of the sump and find a disproportionately high solids and metal content. It is useful to sample from the sump to see what your wastewater contains, but the samples you obtain may mislead you because the sump tends to concentrate solids over time.

If you put a pan or bucket under the part(s) being rinsed, you collect rinsewater which has not been mixed with wastewater that may have changed in character. The wastewater will be more representative of what you will be treating. The biggest disadvantage of this method is that you may only be able to collect a portion of the rinsewater. You may collect the dirtiest or the cleanest fraction from the rinsing step and not get a representative sample. Try to collect a sample that contains rinsewater from the initial rinse which will likely contain the highest concentration of process chemicals and contaminants as well as rinsewater from the end of the rinsing which will contain less contamination. If possible, collect it all and mix it thoroughly before taking a sample. Otherwise, use your judgment and get a good blend of rinsewater.

Use an Automatic Sampler

An automatic sampler takes samples at predetermined intervals over a selected time period. They are available from several manufacturers and many distributors. However, a small business may be discouraged from obtaining one because of the price ($1000 to 2500). If you manage a larger business that generates wastewater of particular concern to your pretreatment authority, such as a metal finisher, an automatic sampler might be a good investment. Another option is to rent an automatic sampler to use for your initial wastewater characterization.

An automatic sampler is ideal for obtaining representative samples from an operation that varies throughout the day. An installed sampler is shown in Figure 4. The sampler can be set up to take a sample every 15 min or every 500 gal, for example. You can select a sampler that puts one or more samples into a series of bottles (a sequential sampler) so that you can analyze the samples from discrete time or flow volume periods. Or you can select a sampler that puts all of the samples into one bottle (a composite sampler), thus averaging the wastewater characteristics over the entire sampling period.

The type of sampler selected, sequential or composite, depends on how you intend to treat the wastewater and on regulatory requirements. If you plan to put in a batch treatment system you can use a composite sampler. If you intend to put in a continuous system then you should consider a sequential sampler so that you can determine how much the wastewater contaminant concentrations vary over a shift. If the wastewater contaminant concentrations

Figure 4 A typical automatic sampler installation in a small shop.

vary widely, you should consider equalizing the wastewater contaminant concentrations by putting in a surge tank. Further details can be found in Chapter 5.

A typical automatic sampler can be used to sample metals and BOD. Metals are a conservative pollutant. They will not change into other metals, unless they are radioactive, and they will not evaporate significantly from the wastewater under normal conditions. Therefore, they can be collected into an open bottle and held for a time before a sample is split out. Wastewater from food processors tends to be highly variable in BOD. In order to get a representative sample, a composite sample must be obtained. In most cases the sample is flow proportioned so that a sample is taken every so many gallons. This gives an accurate average wastewater concentration for the sampling period.

A typical automatic sampler cannot be used to sample volatile organics or FOG according to EPA sampling regulations. Volatile organics can leave the solution and FOG can stick on the sampling lines and in the sample bottles. Grab samples must be collected. In the case of volatile organics a special bottle must be sealed immediately after it is filled. However, you can use an automatic sampler to get an indication of the volatile organics concentration in your wastewater. The data you collect can help verify if you have a problem with volatile organics in your wastewater. It can help you see if a process is introducing volatile organics and the results of modifying the process. In this regard automatic sampling is a useful diagnostic and process monitoring tool and can help you to stay in compliance, even if you have to take grab samples for regulatory purposes.

Flow-proportioned composite samples will be more representative of the wastewater than time-based samples if the wastewater flow varies. However, setting up a flow-based sampler is more involved than setting up a time-based sampler. A means of flow determination is needed for flow-based samples.

Time-based sampling is adequate in a number of common situations, including processes such as hose rinsing, in which the water is turned on and off frequently, and its flow is the same whenever it is on; processes in which wastewater is moved with a pump which operates at a uniform flow rate; or processes with a rinse tank in which the water flow is uniform. In each of these cases the sample line should be installed in a location that can drain when no water is running.

The line can be placed in the drain pipe if the water flow is deep enough when running to cover the hose inlet. The line can be pushed into a sink trap which will be quickly emptied by the sampler when no water is going down the drain. Or a plastic dam can be placed in the sink drain to puddle water in the sink covering the sample line when water is running in the sink. The dam should be designed to allow the sink to drain when no water flows. In these situations wastewater is collected only when it is running. Therefore, the sample collected will be representative of the average wastewater.

You should not use an automatic sampler to take time-measured samples from a sump, unless the sump is designed to drain when no water is flowing into it. If you stick the sampling line into a sump that does not drain, then even if no water is flowing into the sump you are collecting a sample. This can seriously affect the results of the sampling. For example, you may run a lot of clean water through the sump in a relatively short period, then run some dirty water through the sump, and then not run any more water through the sump for the rest of the day. The sampler will pick up the dirty water for a longer time than it picked up the cleaner water, even though more clean water was run through. Therefore, the average contaminant concentrations will be biased high.

Collect All of the Wastewater

Many small shops generate from 1 to 250 gal of wastewater per day. This volume of water can be practically collected and batch treated by a small shop. The best way to characterize your wastewater in this case is to collect a batch of the wastewater, mix it uniformly, and take a sample. Collect the wastewater in a clean sump or in a tank or drum. The sample will be representative of what you will be treating. Businesses with large water flows and tanks also can use this method. The author has collected and sampled batches of wastewater ranging from 1 to 50,000 gal.

Take Grab Samples Manually at Predetermined Intervals

If you cannot collect all of the wastewater and cannot get an automatic sampler, you can sample at intervals by hand. Dip a sample out of the drain line, put a bottle under the discharge, or use a bucket or pan. Sample over the range of conditions normally present in the process line. Have the sampler analyzed individually or mixed together as a composite depending on the data needs.

How Will the Samples Be Handled?

The sample handling requirements depend on the purpose of the sample analysis. Some samples must be taken in special bottles, some samples must be preserved by acidification, and others need to be refrigerated. The length of time the samples can be held before analysis also depends on the nature of the analysis. Check with your laboratory and your pretreatment authority to determine how to preserve the samples and how long the sample can sit before analysis, to ensure that the analytical results are valid.

What Methods Will Be Used to Analyze the Samples?

Analytical methods are established for each wastewater parameter. EPA SW-846 and Standard Methods specify acceptable methods. Make sure that the laboratory you are using follows the specified procedures. Your laboratory can help you to decide what analytical protocol to use if you tell them how you are using the analytical results and what the sample contains. This information will also help them to analyze the sample, and to prepare the sample for analysis most efficiently. For example, it can be diluted, if need be, to avoid excessive contamination of their equipment. The laboratory staff can also take the appropriate safety precautions.

After the laboratory has analyzed your first set of samples, review the results with them. Make sure you are handling the samples correctly between the time you take them and when you submit them to the laboratory. Ask the chemists about any problems they had doing the analysis and if they need more information about your wastewater to help them do a better job for you.

The regulations do allow the procedures to be modified if equivalent results are produced and the procedure is acceptable to the pretreatment authority. Check with your pretreatment authority to ensure that the analytical methods you are using correspond with the methods that they accept.

What Information and Data Will Be Reported?

Check with your pretreatment authority to determine their data reporting requirements. They may want laboratory QA/QC (quality assurance/quality control) data which include the results of sample blanks and duplicates. They may also require that the analytical laboratory be accredited or certified by a specified agency. You may be required to provide the data in a specified format or on a specified form. After your first round of sampling, sit down with your pretreatment authority representative and show the data you have and how you intend to report them. Determine at this point, before you invest your time and money, that you are collecting data that are relevant to the pretreatment authority representative and adjust your sampling plan if needed.

3 DETERMINING WASTEWATER MANAGEMENT LIMITATIONS

INTRODUCTION

The chapter summarizes the potential regulations, process concerns, and economic considerations that define the boundaries of wastewater management decisions. Key wastewater management considerations are given in Table 9.

Before you discharge wastewater you should know whether it is legal to do so. Your wastewater must meet discharge limitation requirements, and you should be able to demonstrate to a regulator that it does. It is your responsibility to understand and to comply with the applicable regulations. The regulatory summary is not intended to be comprehensive because wastewater regulations vary between jurisdictions. The intent of this chapter is to give you an idea of what regulations may apply to your situation in the U.S.

The characteristics of the waste must be compared to the discharge limitations, if the waste is to be sewered; the air pollution control requirements, if the waste is emitted to the atmosphere; or the disposal criteria, if the waste is to be landfilled or incinerated. If the waste is to be recycled, it must be acceptable to the recycling company.

You must determine what the costs of wastewater management are and how the costs fit into your budget. If you change your process to reduce the amount of wastewater generated or change the chemicals used in the process, then you need to understand how the changes affect the workflow in your shop and the quality of your products.

REGULATORY CONSIDERATIONS

Wastewater discharge, hazardous waste, air pollution control, solid waste, and industrial safety and hygiene regulations vary from region to region, state to state, and within a given state, depending on local jurisdictions. Regulations may vary even in different parts of a county. Providing specific guidance for

Table 9 Key Wastewater Management Considerations

Know your limitations
Understand the applicable regulations
- Wastewater discharge
- Hazardous waste
- Solid waste
- Air pollution control
- Industrial hygiene and safety

Determine where the wastewater will be discharged
- Sanitary sewer
- Storm drain
- Septic system

Contact your local wastewater pretreatment authority to determine their specific discharge limitations
Know what your recycler requires
Know your waste hauler and offsite treatment facility requirements
Figure out how much you can spend on wastewater management
Determine the impacts on your shop's operation

every locale in the country is beyond the scope of this text. Check with your state and local authorities to determine their requirements. One of the regulatory references in Chapter 9 may help you get started.

Typically, regulations pertaining to wastewater, hazardous waste, air emissions, solid waste, and industrial safety and hygiene are enforced by different agencies and by different sections within a given agency. You should determine who to talk to about each regulation. A contact in one agency may be able to give you a lead for a contact in another agency which deals with a different regulation. Make a list of all of the agencies which may have jurisdiction over your wastewater management and establish a contact in each of them.

Agency staff should try to be aware of the requirements of other agencies and try to ensure that their regulations do not conflict. However, because environmental regulations are complex and often overlap, conflicts are inevitable. You must try to comply with all of the regulations. Agency inspectors tend to focus on their own requirements and may not always consider the requirements of other agencies. One of the best ways to cope with the uncertainties is to become familiar with the requirements of each agency and ask your contacts at the agencies to help resolve regulatory conflicts.

Agencies are often willing to collaborate on inspections and visit shops together. They can discuss how their various requirements affect your shop while they are at your site. This is helpful because you can step out of the middle a bit and point out the difficulties in complying with conflicting regulations, instead of trying to mediate between agencies. The involved agencies can try to work out a compromise which is satisfactory to them and possible for you to comply with.

The environmental regulations are numerous. The intent of the regulations is to protect public health by maintaining or improving environmental quality. This text focuses on the regulations applicable to wastewater management. The following are some regulations that must be considered:

Wastewater discharge
Solid waste disposal
Hazardous waste disposal
Air pollution control
Industrial hygiene and safety

Wastewater Discharge Regulations

The Federal Clean Water Act of 1977 (CWA) mandates the regulation of the discharge of pollutants from nonpoint sources (such as farm or highway runoff), point sources (your shop sewer pipe effluent), or contaminated storm waters (your parking lot and roof drain rainwater) to any waters of the U.S. The CWA directs the U.S. EPA to regulate industrial dischargers and municipal treatment plants. The EPA responded with a regulation known as 40 CFR Subchapter N – Effluent Guidelines and Standards. 40 CFR gives EPA the authority to enforce the National Pollution Discharge Elimination System (NPDES), which requires that all industrial and municipal wastewater and contact storm water comply with permit limitations. The EPA also promulgated regulations which gives it the authority to enforce the Nonpoint Source Program (NPS), which requires that states control pollution from nonpoint sources. This text focuses on point source industrial wastewater discharges. It does not directly address nonpoint sources or storm water.

The programs establish discharge limitations for a number of pollutants, including conventional, toxic, and nonconventional pollutants. Conventional pollutants include biological oxygen demand (BOD), pH, total suspended solids (TSS), fecal coliform, and fats, oil, and grease (FOG). Toxic pollutants include metals (such as arsenic, cadmium, chromium, copper, lead, nickel, silver, and zinc) and toxic organics. Toxic organics include several broad classes of organics including volatile organics (such as 1,1,1-trichloroethane, benzene, ethylbenzene, methylene chloride, toluene, and trichloroethylene), semivolatile organics (such as di-*n*-butyl phthalate, naphthalene, *p*-chloro-*m*-cresol, and phenol), pesticides (such as DDT, dieldrin and heptachlor), and polychlorinated biphenyls (PCBs). Specific toxic pollutants are called out in Appendix B to 40 CFR 65 (see Table 10). Nonconventional pollutants include ammonium and phosphate.

The EPA has authorized most states to manage the NPDES program for dischargers in their jurisdiction. In some states the EPA directly manages the NPDES program. The EPA expects that all states manage the NPS program and retains oversight of the state programs.

You should talk to your local regulators before you install your wastewater treatment system. The discharge of your industrial wastewater to the sanitary sewer, storm sewer, or septic system may be restricted. They can help you to understand the regulations and help to ensure that your system complies. They have thoroughly evaluated how to apply the various applicable regulations in the context of your discharge requirements. Take advantage of what they

Table 10 Toxic Pollutants Called Out in Appendix B to 40 CFR 65

- Acenaphthalene
- Acrolein
- Acrylonitrile
- Aldrin/dieldrin
- Antimony and compounds
- Arsenic and compounds
- Asbestos
- Benzene
- Benzidine
- Beryllium and compounds
- Cadmium and compounds
- Carbon tetrachloride
- Chlordane (technical mixture and metabolites)
- Chlorinated benzenes (other than dichlorobenzenes)
- Chlorinated ethanes (including 1,2-dichloroethane, 1,1,1-trichloroethane, and hexachloroethane)
- Chloroalkyl ethers (chloroethyl and mixed ethers)
- Chlorinated naphthalene
- Chlorinated phenols (other than those listed elsewhere; includes trichlorophenols and chlorinated cresols)
- Chloroform
- 2-Chlorophenol
- Chromium and compounds
- Copper and compounds
- Cyanides
- DDT and metabolites
- Dichlorobenzenes (1,2-, 1,3-, and 1,4-dichlorobenzenes)
- Dichlorobenzidine
- Dichloroethylenes (1,1- and 1,2-dichloroethylene)
- 2,4-Dichlorophenol
- Dichloropropane and dichloropropene 2,4-dimethylphenol
- Dinitrotoluene
- Diphenylhydrazine
- Endosulfan and metabolites
- Endrin and metabolites
- Ethylbenzene
- Fluoranthene
- Haloethers (other than those listed elsewhere; includes chlorophenylphenyl ethers, bromophenylphenyl ether, bis- (dichloroisopropyl) ether, bis-(chloroethoxy) methane, and polychlorinated diphenyl ethers)
- Halomethanes (other than those listed elsewhere; includes methylene chloride, methylchloride, methylbromide, bromoform, dichlorobromomethane)
- Heptachlor and metabolites
- Hexachlorobutadiene
- Hexachlorocyclohexane
- Hexachloropentadiene
- Isophorone
- Lead and compounds
- Mercury and compounds
- Naphthalene
- Nickel and compounds
- Nitrophenols (including 2,4-dinitrophenol, dinitrocresol)
- Nitrosamines
- Pentachlorophenol
- Phenol
- Phthalate esters

Table 10 Toxic Pollutants Called Out in Appendix B to 40 CFR 65 *(continued)*

- PCBs
- Polynuclear aromatic hydrocarbons (including benzanthracenes, benzopyrenes, benzofluoranthene, chrysenes, dibenzanthracenes, and indenopyrenes)
- Selenium and compounds
- Silver and compounds
- 2,3,7,8-Tetrachlorodibenzo-*p*-dioxin
- Tetrachloroethylene
- Thallium and compounds
- Toluene
- Toxaphene
- Trichloroethylene
- Vinyl chloride
- Zinc and compounds

know. They are usually willing to work with you to increase awareness of environmental concerns and regulations and to help you comply with their requirements.

Sanitary Sewers

Sanitary sewers convey wastewater to a municipal treatment plant where the wastewater is treated before it is discharged to surface water. Municipal treatment plants are required to obtain an NPDES permit and comply with its limitations. The permit limitations depend on the conditions specific to the permitted facility. In order to comply with their permit conditions they must regulate industrial discharges. The state may delegate regulatory authority to a local industrial waste pretreatment program. Most large and many midsized local governments (such as cities or counties) have an industrial waste pretreatment program. Many small cities and sparsely populated counties leave it to their state to regulate industrial discharges.

If you discharge industrial wastewater or contaminated storm water to a sanitary sewer you are an indirect discharger subject to industrial pretreatment permitting requirements. You may be required to obtain a discharge permit or authorization depending on the nature of your discharge, or you may be subject to compliance with best management practices (BMPs) which call out acceptable waste management procedures for specific processes and waste.

A typical municipal treatment plant is designed to use biological treatment and settling to remove BOD and TSS from the wastewater. Some constituents of industrial wastewater are compatible with a municipal wastewater collection and treatment system and can be handled by the facility. Some are not, and can pass through the facility, causing adverse downstream environmental impact, or can upset the biological activity, causing a treatment plant upset and resulting in a permit violation. Others can create hazardous conditions in the sewers or damage equipment. Therefore, municipal treatment plants regulate industrial discharges.

Table 11 Summary of Federal Industrial Waste Discharge Prohibitions

- Pollutants which cause pass-through or interference with the POTW
- Pollutants which create a fire or explosion hazard in the POTW collection and treatment system, including waste streams with a flash point of <140°F
- Pollutants which will cause corrosive structural damage to the collection and treatment system, but in no case discharges with a pH of <5.0, unless the POTW can specifically accommodate such discharges
- Solid or viscous pollutants in amounts which will cause obstruction to the flow in the POTW collection or treatment system, resulting in interference with operations
- Any pollutant, including oxygen-demanding pollutants (BOD, etc.), released at a flow rate and/or pollutant concentration which will cause interference
- Heat in amounts which will inhibit biological activity in the treatment plant, resulting in interference, but in no cases heat in such quantities that the temperature at the treatment plant exceeds 104°F, unless the POTW has alternative temperature limits
- Petroleum oil, nonbiodegradable cutting oil, or products of mineral oil origin in amounts that will cause interference or pass-through
- Pollutants which result in the presence of toxic gases, vapors, or fumes within a POTW in a quantity that may cause acute worker health and safety problems
- Any trucked or hauled pollutants, except at discharge points designated by the POTW

The EPA has established national pretreatment standards for indirect dischargers. Prohibited discharges are identified in 40 CFR 403.5. These prohibitions, which apply to any industrial wastewater discharge, are summarized in Table 11. This table is presented to give you an idea of the basis of the limitations implemented by local pretreatment authorities.

The EPA directs each POTW with a pretreatment program to develop and enforce specific limits to implement these discharge prohibitions. It directs all other POTWs to develop and enforce specific limits to ensure compliance with the POTWs NPDES permit and sludge use or disposal practices. These specific discharge limitations, known as "local limits", are established by the local permitting authority and are deemed pretreatment standards for the purposes of the CWA [40 CFR 403.5 (3)], giving the local agency the authority to enforce the local limits.

The regulations state that specific effluent limits shall not be developed and enforced without individual notice to persons or groups who have requested such notice and without giving them an opportunity to respond. In other words, the POTW must give the public, including regulated businesses, an opportunity to review and respond to proposed limitations. Therefore, you or your trade association may want to keep in touch with your pretreatment regulators so that you are aware of upcoming regulatory changes. Talk with other businesses and find out who is active. Find out how you can contribute to the regulations.

The local limits may be more stringent than federal wastewater discharge limitations, but they cannot be less stringent. Local limit development results in regulating the amount of pollutants the treatment plant can accept, while protecting the health and safety of collection and treatment plant workers and the public, producing effluent and sludge of acceptable quality, protecting the integrity of the collection and treatment system, and complying with hazardous waste and air pollution control regulations. The local limits that you must comply with are specific to your sewer district.

A summary of the local limits developed by the Washington State King County Department of Metropolitan Services (Metro) is given in Table 12, as an example. The local limits include the general federal pretreatment limitations and other limitations specific to Metro's conditions. This table is presented to give you an idea of what your local limits may look like.

The EPA has also developed categorical discharge limitations for specific industrial operations (40 CFR). Table 13 lists those operations. If your facility produces wastewater from operations included in this list, then, in general, you are required to obtain a wastewater discharge permit. Certain exceptions (see Appendix D to Part 403–Selected Industrial Subcategories) should be discussed with your local authority that issues the permit. Categorical dischargers are subject to both federal categorical limitations and local limits. They are issued a permit which includes a combination of the most stringent limitations of both.

Periodically, the EPA publishes pretreatment standards for additional industrial categories. It is the job of your local pretreatment authority to keep current with the changing federal regulations. Ask them if you are in doubt.

If you discharge to a sanitary sewer going to a municipal treatment plant, the first agency to call is the pretreatment authority which works with the treatment plant. To find out who to talk to, contact the utility that bills you for sewer or water service. Sewer usage fees, which include industrial waste surcharges, are often collected by the sewer district. Ask them who is responsible for industrial waste pretreatment, or contact your state industrial wastewater pretreatment coordinator. State pretreatment coordinators are aware of the various local pretreatment jurisdictions in their state.

When you talk to your pretreatment authority, be prepared to tell them how your wastewater is generated, how much you intend to discharge, and what the wastewater contains. They will give you guidance on how to proceed. If the discharge is obviously acceptable to the sanitary sewer, they may give you verbal approval to discharge or a letter of permission. If the discharge is questionable or obviously requires a permit, they will request that you submit your wastewater characterization in the form of a discharge permit application. If your wastewater must be pretreated to meet their permit limits, they will ask for a report documenting how you intend to treat the wastewater. Preparation of such a report is covered in Chapter 7.

A wastewater discharge permit will usually include a compliance monitoring schedule specifying the minimum number of samples to be taken and the parameters to be analyzed. The permit will also specify the sampling site. A permitted business will be sampled and inspected periodically by the permitting agency.

Apply for a permit well in advance of when you will need it. Even if the process goes smoothly and you provide all the information required in a timely manner, you may not receive your permit until months after you have applied

Table 12 Example Industrial Wastewater Discharge Local Limits

- No pollutant that creates a fire or explosion hazard in any sewer or treatment works, including but not limited to waste streams with a flash point of <140°F
- No pollutant causing, at the point of discharge or at any point in the system, two successive readings on an explosion hazard meter of >5% nor one reading >10% of the lower explosive limit as read by the meter
- Pollutants subject to the fire or explosion hazard limit include, but are not limited to, gasoline, kerosene, naphtha, benzene, toluene, xylene, ethers, alcohols, ketones, aldehydes, peroxides, chlorates, perchlorates, bromates, carbides, hydrides, and sulfides, and any other substance that Metro, a fire department, the state, or the EPA have notified the discharger are a fire hazard or hazard to the system
- The settleable solids concentration must be <7.0 ml/L
- No organic pollutant discharges that result in the presence of toxic gases, vapors, or fumes within a public or private sewer or treatment works in a quantity that may cause acute worker health and safety problems: the organic pollutants include, but are not limited to, any organic compound listed in the 40 CFR Section 433.11 Total Toxic Organics definition, acetone, 2-butanone, 4-methyl-2-pentanone, and xylenes
- Dischargers are required to implement housekeeping and BMPs in order to prevent the discharge of a concentrated form of any of the above organic pollutants
- Individual permit limits for specific industrial discharges may be established for the above organic pollutants on a case-by-case basis
- The atmospheric hydrogen sulfide concentration must not exceed 10.0 ppm at a designated monitoring manhole
- Soluble sulfide concentrations may be established on a case-by-case basis
- No discharge with a single sample pH of <5.0 or a composite average of 4 samples of pH <5.5 or a 15-min recording average of pH <5.5
- No process rinsewater with pH >12.0
- No discharges of >50 gal/day of caustic solutions equivalent >5% NaOH by weight or pH >12.0, unless authorized by permit
- No nonpolar (petroleum) FOG concentrations exceeding 100 mg/L

No metal and cyanide discharges exceeding the following limitations:

Parameter	All IUs and SIUs Instantaneous max (ppm)	IUs >5000 gpd and SIUs Daily avg. (ppm)	IUs <5000 gpd Daily avg. (ppm)
Arsenic	4.0	1.0	4.0
Cadmium	0.6	0.5	0.6
Chromium	5.0	2.75	5.0
Copper	8.0	8.0	8.0
Lead	4.0	2.0	4.0
Mercury	0.2	0.1	0.2
Nickel	5.0	2.5	5.0
Silver	3.0	1.0	3.0
Zinc	10.0	5.0	10.0
Cyanide	3.0	2.0	3.0

IUs = Industrial users which discharge wastewater to the sanitary sewer; SIUs = Significant industrial users, as defined in 40 CFR 403.3; includes federal categorical dischargers and all dischargers that have a reasonable potential for adversely affecting treatment collection and treatment operations.

- Individual permit limits for specific companies may be established for compounds not specifically listed or for listed compounds at levels higher or lower than the above limits, dependent upon a case-by-case evaluation
- In addition to concentration limits, permit limits may also include mass limits stated as total pounds of a pollutant allowed per day

Table 13 Federal Categorical Industries

Aluminum Forming	Metal Finishing
Asbestos Manufacturing	Metal Molding and Casting
Battery Manufacturing	Nonferrous Metals Forming
Builder's Paper	Nonferrous Metals Manufacturing
Carbon Black	Paint Formulating
Cement Manufacturing	Paving and Roofing (Tars and Asphalt)
Coil Coating and Can Making	Pesticides
Copper Forming	Petroleum Refining
Electrical and Electronic Components	Pharmaceuticals
Electroplating	Phosphate Manufacture
Feedlots	Porcelain Enameling
Ferroalloy Manufacturing	Pulp and Paper
Fertilizer Manufacturing	Rubber Processing
Fruits and Vegetables Processing	Seafood Processing
Glass Manufacturing	Soaps and Detergents Manufacturing
Grain Mills Manufacturing	Steam Electric
Ink Formulating	Sugar Processing
Inorganic Chemicals Manufacturing	Timber Products
Iron and Steel Manufacturing	Plastics Molding and Forming
Leather Tanning and Finishing	Textile Mills
Meat Processing	

for it. For example, categorical dischargers according to federal regulations must give 90 day notice prior to beginning discharge. A local authority may require 60 days notice for other permitted discharges. The clock does not start ticking with your first contact to the regulatory authority. The pretreatment authority probably will not consider a phone call or letter asking for information about permitting requirements to constitute notification; this will occur when you submit a completed permit application.

Many small businesses produce wastewater which does not present a large risk to a municipal treatment plant. Therefore, a pretreatment program may allow the discharge of industrial wastewater without a formal permit and discharge monitoring requirements, if a business follows BMPs. In these cases the pretreatment authority has adequate knowledge of the waste based upon the nature of the business practices.

The BMPs are developed based on the type of business or waste-generating process to provide guidance for the proper management of wastewater and the processes that contribute pollutants to the wastewater. The BMPs approach is used to ensure that the municipal treatment meets its requirements without imposing the burden of permit compliance on small dischargers. The approach is useful for groups of businesses which generate small amounts of similar wastewater which typically meet local discharge limitations without pretreatment, or would meet local limits if certain wastes were kept out of the discharge. An example wastewater BMP for screen printing wash water is summarized in Table 14.

Your local pretreatment authority may have a special program to deal with moderate-risk wastes. The program may work specifically with small business

Table 14 Example Wastewater BMPs

Keep excess ink out of the sewer
- Remove as much ink as practical from a screen before solvent or water cleaning
- It is acceptable to rinse a small amount of residual ink to the sewer
- Use inks which do not contain regulated metals when possible

Keep solvent out of the sewer
- If solvent is used to clean the screens, then dry the screen before taking it to the sink for reclamation
- Do not solvent clean screens in a sink that drains to the sewer; use a separate solvent cleaning station
- Use alternatives to volatile and chlorinated solvents when possible

Consider worker health and safety, fire, air pollution, wastewater, and hazardous waste regulations when changing or implementing a process

to help them comply with regulations, including wastewater discharge regulations. Find out if this is the case in your area. These programs typically do not have enforcement power, but work to educate small businesses without automatically fining them. They take a proactive approach and make it a point to try to understand the problems faced by small businesses trying to comply with complex environmental regulations. They may know of another business, similar to yours, which has found a workable solution to an environmental problem you have, or they may be able to get you in touch with a trade organization that can help. They also may have sampled and characterized wastewater from businesses similar to yours and be able to offer advice on wastewater management.

Storm Drains

If you discharge industrial wastewater or contaminated stormwater directly into surface waters you are required to obtain an NPDES permit as a direct discharger. If your wastewater runs onto the ground, into a drain that discharges on or under the ground, or into a lake, stream, river, or other body of water without going through a municipal treatment plant, then you are discharging directly to surface waters. Contact your regional EPA office to determine whether they have jurisdiction at your location or have delegated authority to your state, or call your state environmental department directly to determine who to talk with.

The NPDES direct discharge permit requirements are typically much more stringent than municipal treatment plant discharge requirements. The storm water discharge limitations with which you must comply depend on your state's program. Several permitting approaches are used, including individual permits, group permits, and general permits.

The NPDES permit limits may be developed specifically for the discharge under review. The discharger must provide wastewater characteristics as well as information about the receiving water so that permit limits can be developed. The limits are based on EPA Water Quality Criteria. An individual permit typically requires more agency resources to develop than any of the other

Table 15 Summary of Federal Storm Water Discharge Permit Requirements as Interpreted by Washington State

Storm water discharges must not cause a violation of surface water, groundwater, or sediment quality standards
Discharges to a storm sewer or surface water of process or noncontact cooling water are prohibited, unless covered by an NPDES permit
Every permitted facility must develop and implement a storm water pollution prevention plan which identifies potential sources of pollution and describes the practices used to control them
Best Pollutant Control Technology (BCT) will be applied for conventional pollutants and Best Available Technology Economically Achievable (BAT) for toxic and unconventional pollutants
All industrial permit holders must conduct two annual inspections
A dry season inspection must determine if any nonstorm water discharges exist
Any nonstorm water discharges must be eliminated or covered by an NPDES permit
A wet season inspection must verify that the description of potential pollution sources is accurate, that the site map reflects current conditions, and that storm water controls are adequate
The permit holder must make an assessment of the potential of the storm water discharge to violate surface water, groundwater, or sediment quality standards; permit holders discharges determined to have a high potential for violating standards will be required to monitor their discharges
The permit does not require monitoring of discharges; however, monitoring is encouraged
Discharges of storm water to a sanitary sewer must be permitted by the local pretreatment authority

permits, and agencies are struggling to implement the stormwater discharge program. Therefore, individual permits are difficult to obtain.

Almost all states have developed a general permitting program. The permits are developed for industry classifications that generate similar wastewater. For example, Washington State has developed a general permit for boatyards, shipyards, and marinas which includes BMPs. A summary of the EPA NPDES general storm water permit requirements, as interpreted by Washington State, is given in Table 15. Keep in mind that permit requirements can vary widely. A general permit can facilitate the permitting process because permit requirements are the same for all participating businesses and not developed on a case-by-case basis. To get a general permit you will usually file a notice of intent. At that point you will be expected to comply with the general permit conditions. You may not be inspected for a while, but when you are inspected you will need to demonstrate compliance.

Some states have a group permitting alternative. A group permit is developed for multiple sources which file for a permit together. The permitting requirements are usually more strict than the general permitting requirements. Also, you may go through the work of submitting a group permit application and then be told to use a general permit anyway. Compare the requirements of the group and general permits and determine which type of permit will work best for your situation.

Housekeeping, storm water management, monitoring, and recordkeeping requirements will depend on your permit conditions regardless of the type of permit you get. Some of the requirements may take some study for you to understand. For example, the compliance sampling plan may specify the use of specialized equipment and adherence to a rigorous procedure. You may need help to get started correctly or to take all of the samples. How much you want to do yourself is up to you. Contact pollution control managers at businesses similar to yours to find out how far along they are, and contact your trade organization. Get some insight into how to demonstrate permit compliance. Do this before you set out on your own. It can save you time and frustration.

Septic Tank Systems

If your wastewater drains to a septic tank system, also known as an "on-site treatment system", call your state or local on-site waste treatment coordinator to determine the requirements. The coordinator may be a local or state official, depending on who has jurisdiction in your area. Call your local health department. They have jurisdiction over septic tank systems and can refer you to the appropriate agency.

An on-site treatment system is designed to treat domestic wastewater. Domestic wastewater includes water from ordinary bathroom, kitchen, and laundry room practices. Wastewater generated by other practices should not be discharged to an on-site treatment system unless it has characteristics similar to domestic wastewater or is demonstrated to be compatible with the system.

Wastewater which is not compatible can cause the system to fail and produce unacceptable effluent, resulting in contamination of the drain field and downstream water supplies. Cleanup costs can be substantial. The septic tank sludge also may be contaminated and may not be acceptable to a septage hauler, resulting in higher sludge disposal costs. You could also cause the sludge to be designated as hazardous waste, making its disposal difficult and expensive.

If used for the disposal of hazardous waste, an on-site treatment system may be considered an underground injection well under the Resource Conservation and Recovery Act (RCRA). It is very unlikely that you could get a RCRA permit to use an on-site treatment system to dispose of hazardous waste. Even if you could, it would be very expensive and probably not cost effective.

If you are using an on-site treatment system to dispose of your industrial wastewater, carefully consider whether it is an appropriate disposal method. You are liable for your discharge whether or not you understand the regulations. It is your responsibility to find out what regulations apply and to comply with them.

Hazardous Waste

The EPA has defined solid and hazardous waste in the RCRA regulations (40 CFR 261). The process of designating a hazardous waste (determining if your waste is indeed a hazardous waste) is complex. Refer to other texts for guidance on this issue.

Your wastewater may be hazardous waste. Hazardous waste should not be discharged to a septic system or storm sewer without specific authorization and is strictly regulated when discharged to a sanitary sewer. It is important for you to understand how your local authorities regulate hazardous waste. Hazardous waste regulations vary between states. Contact your state's hazardous or moderate-risk waste program (environmental department) for more specific information.

States have set up moderate-risk waste programs to help small businesses understand and comply with hazardous waste regulations. Moderate-risk waste is hazardous waste generated in quantities that do not trigger full RCRA regulatory requirements. Most moderate risk waste programs take an cooperative educational approach, rather than an enforcement penalty approach. They have found that businesses tend to respond better to someone who has come to help than to someone who has come to write them up for noncompliance. Each moderate-risk waste program is unique and has different resources. Give your program a chance and see what it can do for you.

Federal pretreatment regulations allow what is called a Domestic Sewage Exclusion (DSE) for industrial wastewater. In summary, the DSE excludes industrial wastewater from regulation as a hazardous waste. Once industrial wastewater is discharged to a sewer conveying domestic wastewater, it is regulated as domestic wastewater. This has significant implications for a municipal treatment plant. Without the exclusion, the treatment plant sludge might be considered to be hazardous waste because it contains hazardous waste regulated under RCRA. With the DSE, the sludge is not considered hazardous waste unless it tests out as hazardous waste.

States have produced their own interpretations of the DSE, and the exclusion is handled differently by various pretreatment programs. Hazardous waste and sewer discharge regulations intersect at the DSE, and regulators are typically very cautious in its interpretation. The issue is complicated by hazardous waste disposal reporting requirements. Your local pretreatment coordinator can help you determine what hazardous waste is acceptable for discharge to their collection system and what procedure you need to follow to discharge it.

Solid Waste

Solid waste includes what you put into your trash and garbage cans and dumpsters. If you treat wastewater you usually end up with a slurry, sludge, dirty filters, contaminated carbon, and/or other residuals. Do not assume you can throw the waste into your dumpster. Municipal landfills do not want your hazardous waste. Most large solid waste programs have a waste screening section. The section establishes criteria for what wastes they will accept and what information they need to make that determination.

You are responsible for characterizing your waste to determine if it meets the requirements of the municipal landfill. The waste screening staff can tell you what information they need. You must provide it. The waste screeners have seen a variety of waste and therefore may have data on similar waste from another business. This information can help them make a determination about your waste. They may have enough information to accept the waste without additional testing, or, if testing is required, they may be able to reduce the cost of testing by specifying certain tests based on knowledge of what is likely to be of concern in the particular waste.

Before you throw your wastewater treatment residuals in your dumpster, contact your local public health department. They are usually responsible for municipal landfill waste screening. If in doubt, ask your municipal waste hauler who to call.

Your local fire department will also typically show an interest in your solid waste handling practices. They will inspect your garbage cans and dumpsters to determine if you are following fire code requirements.

Whatever you do, do not throw the wastewater treatment residuals on the ground. You will end up having to do a lot of explaining. Besides, it causes an unsightly mess which could keep customers away. To sum up, dispose of your wastewater treatment residuals properly.

Air Pollution Control

The Federal Clean Air Act was enacted to protect and enhance the quality of the air resources of the U.S. so as to promote the public health and welfare and the productive capacity of its population. The Act states that the prevention and control of air pollution at its source is the primary responsibility of states and local governments. The Act establishes or provides for the establishment of ambient air quality criteria for specific pollutants. States must enact air pollution control programs to meet these criteria.

State and local regulations apply to the emission of air pollutants from your facility. Contact your state department of environmental quality or ecology or your local air pollution control agency to determine who will have jurisdiction in your situation.

The agency you deal with will want to know what you are treating and what may be emitted from the waste. For wastewater treatment they will be especially concerned about volatile organics, odors and mist, or vapors. Volatile organics contribute to smog formation and may be toxic or present a cancer risk. The agency will want to know if volatile organics are stripped from the water in the treatment process and, if so, how much. Odors may create a public nuisance and therefore are discouraged. Mist and vapor may carry contaminants which threaten the public health. Your air pollution control authority will tell you what you need to do based on your situation and their regulations.

The requirements will vary between states and between air pollution control districts in each state. Find out what the rules are before you change your industrial process or build your wastewater treatment system. You do not want to have to go back and redo it because it does not meet air pollution control requirements.

Industrial Hygiene and Safety

The Federal Occupational Safety and Health Act (OSHA) was established to assure so far as possible every working man and woman in the U.S. safe and healthful working conditions and to preserve our human resources. It encourages employers and employees in their efforts to reduce the number of occupational safety and health hazards at their places of employment and to institute new programs and to perfect existing programs for providing safe and healthful working conditions. It authorizes the Secretary of Labor to set mandatory occupational safety and health standards, and it provides for an enforcement program which includes a prohibition against giving advanced notice of any inspection. The states are encouraged by OSHA to assume the fullest responsibility for the administration and enforcement of their occupational safety and health laws.

State occupational health and safety programs typically adopt the OSHA standards and apply them to businesses in which an owner has employees. Health and safety standards cover a broad range of potential hazards, including shop air quality (dust, fumes, smoke, and vapor levels), skin exposure (to chemicals), heat, noise, moving equipment, electricity, confined spaces, and falls.

A treatment system can present all of the above hazards. Contact your state occupational health and safety program to determine what their rules are and how they apply to your situation. Find out what safeguards, controls, and procedures you need to use. Develop a health and safety plan for the treatment system and its operation. Think about health and safety from the outset of the design process.

Some states have established educational programs in which a staff member will take a look at your site and your plans, pointing out where you should make improvements. If you are near a university, find out if they have an environmental health program. If so, they may have a program in which students canvass a business to conduct a health and safety audit and make recommendations to improve shop conditions. The students gain valuable experience and you benefit from a thorough audit. Also, talk to pollution control and industrial hygiene and safety managers at other businesses. They may offer useful advice.

RECYCLER'S REQUIREMENTS

A variety of companies, both large and small, accept recyclable material. Some chemical suppliers will collect spent solutions and recondition them for reuse. Some wastewater treatment equipment suppliers will handle the sludge produced by the equipment. Metal-bearing residuals may be accepted by foundries. If possible, recycle your spent material and wastewater treatment residuals.

Recyclers have worked out procedures for characterizing, packaging, and shipping reclaimable material. They are subject to the same regulations that you are. They also have to ensure that they remain in compliance. They must screen carefully the materials they accept. Find out what are their requirements.

To identify recyclers, look in the phone book, ask your trade association and other businesses, contact a materials exchange program, ask your moderate-risk waste program staff, and talk to your chemical supplier.

TREATMENT, STORAGE, AND DISPOSAL REQUIREMENTS

Hazardous waste treatment storage and disposal facilities (TSDs) are strictly regulated and have high visibility. They have stringent monitoring, recordkeeping, and reporting requirements. You must comply with their requirements for them to accept your waste.

If you send waste to a TSD, they will require that it be profiled to determine its regulatory status. A waste profile is a characterization and designation of a waste. Be prepared to provide detailed information on how the waste was generated. This will help the TSD to profile it and can reduce the cost of profiling by narrowing the number of tests to be conducted based on your knowledge of what the waste contains. If you profile the waste yourself make sure that you understand what the TSD wants or they will have to profile it again.

The waste is subject to federal and state transportation regulations. Make sure that you understand them. The TSD should be a licensed waste hauler. If you transport the waste yourself, do it legally. Call your TSD or state department of transportation for advice.

The TSDs offer a variety of services. Some will handle all aspects of your hazardous waste management, including profiling, packaging, and labeling your waste, filling out the necessary paperwork, and transporting and disposing of the waste. However, you are responsible for the waste and the paperwork, regardless of who manages it.

BUDGETARY CONSIDERATIONS

Budgeting is a two-sided issue: (1) the cost of installing, maintaining, and operating a wastewater treatment system, and (2) the amount of money that is available for it. A wastewater treatment system can cost very little to install; e.g., you may be able to build one with existing parts and equipment for a few hundred or a few thousand dollars, or you may spend $50,000 for a skid-mounted ready-to-use 50 gal/min system, or $1 million for a custom-built 100 gal/min system to treat a difficult waste stream. Your budget will also range widely depending on your cash flow and profit margin.

Before you get too far into the design process you should get a rough idea of how much you can spend and how much the various treatment options cost. Implementation of the system includes:

Development
Permitting
Capital equipment procurement
Installation
Training
Operation and maintenance

The costs of each of these steps depends on the circumstances peculiar to your shop and the wastewater you generate. If you have a small volume of easily treated, well-characterized wastewater, your implementation costs will be lower than for a shop with a large volume of complex, hard-to-treat wastewater.

You can do all of the work yourself or get help from equipment suppliers or consultants. You can spend your time or pay for someone else's. In any case, you should stay involved in the process so that you understand the system you are installing. You will probably be responsible for running and maintaining it.

Prepare a work plan for the treatment system development. Include a schedule and budget roughly showing the time and money you think you will spend on the project. The first task in the project is to get more information about the cost of implementing a system and how long it will take. As you get into the project you can refine your estimates. By the time you are ready to buy equipment you should have collected enough information to know realistically how much it will cost to put in and maintain. If you do not know, you should not proceed.

SHOP IMPACT CONSIDERATIONS

How will your waste management activities affect your shop operations? Someone will have to oversee waste management. You should clearly assign the responsibility to someone and make it part of their routine duties. Then give them the time they need to do it.

When you put in wastewater treatment you will have to be more aware of the wastewater you are generating than you were when you just let it run down the drain without treatment. People generating the wastewater may have to change their practices so that the treatment system is protected from upsets. Changing people's habits is a challenge. Plan to spend time teaching shop personnel how to work within the new constraints.

Workflow patterns may change because you must separate processes which generate waste that must be kept out of the wastewater. For example, you may need to find a new area for solvent cleaning instead of using the steam cleaning sump or sink to catch the used solvent. This change may disrupt established work patterns. It may also take more time to do a job if the part must be moved or the cleaning process takes longer.

If you change the generating process, you obviously want to consider part quality requirements. If you do subcontract work your choices of process chemicals and procedures may be limited. You must follow the specifications provided by the contractor. Some businesses are constrained by regulatory requirements and must use specified processes. Determine the specifications that you must adhere to and develop process changes within the bounds of the specifications. If you are having a problem complying with a contractor's requirements and regulatory requirements, meet with the contractor to see if specification changes are possible.

The key to developing a successful wastewater treatment system is to understand the constraints within which you are working. Regulatory concerns are certainly important. The treatment system must produce wastewater that meets discharge regulations. You must comply with health, safety, fire, building, and other regulations, but you must also be able to afford to buy and run the system. Also, the system must not upset the operation of your facility. It should fit in smoothly with your workflow. You should select equipment you can understand well enough to run, maintain, and troubleshoot yourself unless you are prepared to depend on a vendor for help. You should try to get a system with enough flexibility that it can accommodate workload fluctuations and processes changes that you plan to make. Do your homework and put in something that works.

4 DEVELOPING WASTEWATER MANAGEMENT ALTERNATIVES

INTRODUCTION

This chapter deals with the development of wastewater management alternatives. Key topics are presented in Table 16. You should use a systematic approach to develop alternatives. The goal is to logically examine the alternatives and to pick the best one based on objective criteria. You can start off with a general list and a rough idea of costs and benefits and then collect more detailed information on the alternatives that appear to be the best for your shop.

You need to evaluate only alternatives that have a chance of being successful in your shop. List alternatives, collect information, and compare them based on what you know. Start off broadly and narrow your list down.

In the broadest terms you have two alternatives: continue to do things the same way or do something different. You probably have reasons why you cannot continue to do things the way you are now, but changing what you are doing must result in benefits. Therefore, the first basis of alternative evaluation is to compare the alternatives to what you are doing now and choose something better.

The first list of alternatives is broad:

Do nothing
Change your processes to eliminate the generation of out-of-compliance waste water
Collect your wastewater and treat it off-site
Install a wastewater treatment system and treat on-site

If you are in a compliance situation (e.g., you cannot discharge your wastewater to the sewer because it does not meet discharge limitations) the first alternative, "do nothing", is not viable. You need to compare alternative methods of meeting sewer discharge limitations.

Table 16 Key Topics for Wastewater Management Alternative Development

Start with the basic alternatives outline
• Do nothing
• Change your processes and do not produce out-of-compliance wastewater
• Collect your wastewater and have it treated offsite
• Install a wastewater treatment system and treat onsite
Identify applicable process changes and treatment technologies
Investigate system implementation options which include
• Do it yourself
• Purchase it
List the advantages and disadvantages of each alternative
Estimate the costs of each alternative
Document the alternatives you identified
• List the potentially viable alternatives as well as the ones that are not feasible
Take advantage of the various information sources

DO NOT PRODUCE OUT-OF-COMPLIANCE WASTEWATER

Must you really build a wastewater treatment plant? Can you change your shop operations so that you do not produce wastewater at all? Or can you alter your processes and produce wastewater that can be discharged legally without treatment? If the wastewater cannot be discharged because of what is in it and you cannot keep out contaminants you must treat it or have someone else haul it away and treat it for you.

Take a look at the shop process list that you put together (Chapter 2). Where is your wastewater generated and where are contaminants introduced? Do the contaminants come from the parts you are processing or from the chemicals you use, or both?

CHANGE THE PROCESS

Do not produce the wastewater if you do not have to. Not generating the waste in the first place is a top priority of pollution prevention. This is easier said than done because water is commonly used in cleaning processes. Moreover, the trend is to switch from solvents to water-soluble cleaners because of the regulatory pressure to reduce solvent use. However, you should determine if you have to produce wastewater before you think about managing its disposal.

Cleaning up and reusing the process water may be an option. I consider this to be equivalent to wastewater treatment because you will probably have to install some sort of treatment system to clean up the water for reuse. If so, then follow the procedures outlined in this book to develop a water recycling system that works. You may not call it a wastewater treatment system, but it serves the same function, and you can evaluate it the same way. See the wastewater treatment alternatives development section of this chapter for help in wastewater recycling.

If the wastewater cannot be discharged because it contains a process chemical that you are using, consider using a different chemical. Change your process to use chemicals which are more acceptable to the regulators. It is the highest priority in pollution prevention and the step you should consider first. For example, you may be using a volatile solvent, like lacquer thinner or methyl ethyl ketone (MEK), to remove water soluble ink from a printing screen. The ink does not contain regulated metals and most has been removed from the screen before taking it to the screen wash sink for final cleaning. The slight ink residue is acceptable to the sewer. The solvent is not. If you switch to a water-soluble cleaning agent that is acceptable to discharge, then you have solved your problem.

In many cases, the soil on the parts causes the wastewater to be out-of-compliance with discharge regulations. For example, if you are repairing engines they are typically covered with oil and may be coated with lead- or zinc-containing paint. You must remove the oil and may remove the paint to repair or rebuild the engines. Hot tanks or jet washers are commonly used to clean engine blocks. The cleaning solution is retained and used to clean many engines before needing to be replaced. As the solution is used, it collects oil and paint. Some of the solution is carried out on the blocks, and the parts are usually rinsed with water after the hot tank or jet wash. At this stage additional oil and paint may be removed. The oil and paint will likely render your wastewater unacceptable for discharge, regardless of the cleaning chemicals you use.

When you are looking for an alternative cleaner keep in mind that just because the label says it is "drain safe" or "biodegradable" does not mean that it is. Alternative solvents may present different health or safety risks than the risks associated with solvents widely in use. In some cases, little information is available to form a judgment. Ask your pretreatment and industrial safety and hygiene authorities what they know about the alternative before you commit to using it. As new alternatives are developed, these agencies try to determine if the alternatives are more environmentally sound than the solvents in common use. Try not to switch to an alternative that causes other problems.

SEGREGATE THE WASTE STREAMS

Another method of producing acceptable wastewater is to keep the offending process waste out of the wastewater and deal with it separately. In other words, segregate your waste streams. Again, take a look at your processes and the wastes that they produce. List the waste streams that may be acceptable to the sanitary sewer as is. List the waste streams that are not. Consider whether you can keep them separate and what shop modifications it would take to do so.

For example, you may be using inks which require solvents, like MEK or lacquer thinner, for removal from the printing screen. Do not use the solvents over your sink or place solvent wet screens in the sink and rinse them with water. Set up a separate area to solvent clean and dry the screens before taking them to the screen wash sink where water is used to remove the emulsion. Solvent is kept out of the wastewater, and the emulsion removal wastewater can be discharged. To make this change you must set up another process area and alter the workflow through your shop. Write down what modifications you would need to make.

CHANGE THE PARTS

You may be able to change the nature of the parts that you are processing to eliminate regulated components from your wastewater. This is an option only if you have control over the parts. You cannot change the parts if you are repairing or cleaning equipment that people bring in.

In some cases you may be able to subcontract the work to a shop that can efficiently manage the waste generated by making or finishing the part. For example, instead of having a metal plating line that you use infrequently, you could arrange for a plating shop to do your plating for you. Of course, you should check to be sure that the plating shop is set up to deal with the challenge of plating waste management.

COLLECT YOUR WASTEWATER AND HAVE IT HAULED AWAY

You could collect your wastewater and have someone else treat it. If you choose this option, little or no treatment of wastewater will be done in your shop. You will still need to know what is in the wastewater, and to avoid putting incompatible wastes in the water (or pay more to get rid of it).

INSTALL A WASTEWATER TREATMENT SYSTEM

If you cannot stop generating out-of-compliance wastewater and do not want to have it all hauled away, you need to develop a wastewater treatment system (the main topic of this book). Alternatives to installing a wastewater treatment system were included because evaluating those alternatives gives you a context within which to develop a successful treatment system. All of the alternatives have advantages and disadvantages. Listing various alternatives for evaluation gives you different ways of looking at your problem and can ensure that you have taken into account everything you should. The more you know about your wastewater, the better chance you have that your treatment system will work.

To determine what sort of treatment system to install you must first identify applicable treatment technologies. Available treatment technologies and examples of several successful systems are included in this book. However, this book cannot include every available technology or give an example which will fit every shop. For more information, contact shops which produce wastes similar to yours, your trade associations, equipment suppliers, the library and bookstores, and regulatory agencies.

Descriptions of Some Available Treatment Technologies

Treatment methods can be divided into three general classes: physical/chemical, thermal, and biological. Physical/chemical methods make use of the physical and chemical properties of the waste and its constituents to effect a desirable change in the waste. Thermal methods use heat to decompose pollutants. Biological methods use bacteria and other organisms to metabolize waste components, removing them from the waste stream. These treatment methods are described briefly below. Check out the references given in Chapter 9 for more details.

Physical/Chemical Treatment

Physical/chemical treatment methods encompass a wide variety of technologies, including gravity separation, filtration, chemical precipitation, evaporation, oxidation, reduction, air stripping, carbon adsorption, ion exchange, adsorption on other media, electrolytic recovery, and membrane separation.

Gravity Separation

Gravity separation is the simplest form of wastewater treatment. It can be used to remove floating oil or solids which are large enough to settle out. Oil water separators are available to remove liquids that float, and clarifiers are available to remove solids that sink.

Oil Water Separators

Oil water separators were designed to recover oil products which spill or leak from petroleum storage and transfer tanks. An oil water separator works by providing time for the oil to float and baffles to keep the floating oil in the separator. An American Petroleum Institute (API) separator is a baffled vault (Figure 5). A 750-gal unit, which can handle a wastewater flow of 10 to 20 gal/min, costs about $2,000 plus installation. Installation of a 4 × 6 × 6 ft vault requires excavation for the tank and pipes and can be costly, depending on what you have to dig through. It is most cost effective to install an API separator during building construction rather than retrofitting it.

Some separators incorporate coalescing plates or other media which can improve their efficiency by helping small oil drops to touch each other and grow large enough to rise quickly (Figure 6). These separators can be made much smaller for a given flow rate than separators with simple baffles. A 2 × 5 × 3 ft unit can handle 10 gal/min and is small enough to stick in a corner of your shop.

An oil water separator must be cleaned out periodically. Solids build up in the bottom, reducing its effective capacity, and plug the channels, causing wastewater to short circuit the unit. Many an old oil water separator has filled with solids due to neglect. Also, oil collects in the separator. Eventually, the floating oil layer will get deep enough that the oil will run under the baffle and

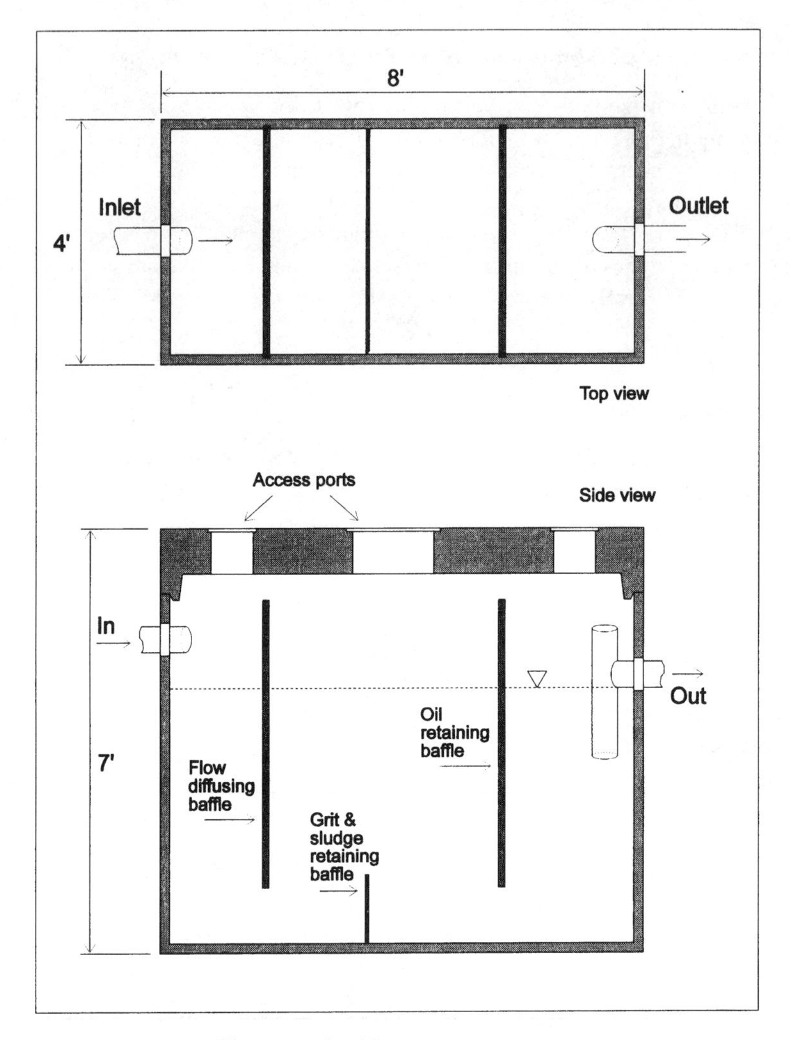

Figure 5 An API oil water separator.

into the sewer. The separator must be regularly maintained. If you have an oil water separator take a look at what is in it and clean it out if needed. Remember to dispose of the contents appropriately.

An oil water separator will not remove emulsified oil. Oil is chemically emulsified by soaps, detergents, and other surfactants. Pumping oily water with an impeller pump can also break up oil drops, forming a mechanical emulsion which complicates oil removal. For example, milk is homogenized by using high-speed impellers which mechanically emulsify the cream. To remove emulsified oil you must break the emulsion, typically by using chemical treatment.

Most oily water produced by small businesses will contain emulsified oil because surfactants are used to remove the oil from dirty parts or equipment.

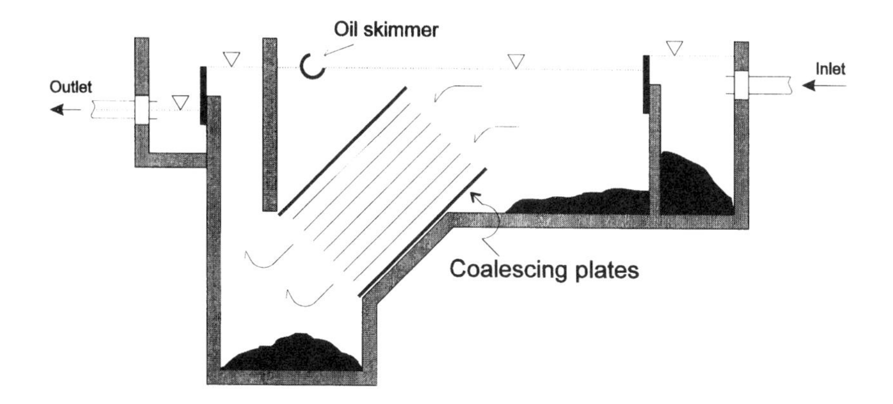

Figure 6 An oil water separator incorporating coalescing plates.

Oily rinsewater from a caustic cleaner will typically contain emulsified oil. Caustic cleaners may have detergents in them, and, even if they do not, hot caustic will turn fats and oils into soap. If your oily wastewater is turbid (that is, cloudy like diluted milk, but probably not white), it probably contains emulsified oil. This is the case with most oily water generated by cleaning and maintenance processes. Therefore, oil water separators are not a good choice for the treatment of the oily wastewater generated by most small businesses.

To get an idea as to whether your oily water is emulsified, take a sample and let it sit for 1 hr. If the oil floats and you are left with water that you can see through, then an oil water separator may produce wastewater that meets your discharge limitations. You can pursue further evaluation of a separator. On the other hand, if the oily water is still turbid, and you cannot read a newspaper through a jar of it, an oil separator most likely will not be effective. You can confirm this by having the sample analyzed for fats, oil, and grease (FOG). A 100 mg/L nonpolar oil and grease limit is probably the highest you will encounter, which is a small amount. If you cannot see through the water, it probably contains >100 mg/L emulsified oil.

You may be able to produce oily wastewater that is treatable by an oil water separator by changing your process. Try a different cleaner, hot water, or steam without chemicals to clean your parts and equipment. See what kind of wastewater you produce. Some shops can get by without using detergents or caustics.

Clarifiers

A clarifier is designed to remove settleable solids. A clarifier is typically used after chemical or biological treatment to remove solids produced by the process. Finely divided or colloidal solids that do not settle are not removed.

Industrial cleaners and other chemicals which are used to make solids "water wettable" tend to create stable suspensions of solid particles. These chemicals are designed to keep the dirt in the water. If the suspended dirt contains enough regulated metals or oil to put your water out-of-compliance, then you must remove the suspension from the water. A clarifier alone will not do it.

To determine whether your wastewater may be amenable to treatment by simple settling, take two samples and let them set for 1 h. Then pour one of them into another sample container, leaving any settled solids behind (in chemistry jargon this is known as "decanting"). Have the water that you decanted (the supernatant) and the sample of the original wastewater analyzed for metals or other regulated substances. If a significant amount of contaminants was removed by settling, and the wastewater is within discharge limits, then you can further evaluate settling. If the solids do not settle effectively, you must look for another treatment method. Consider chemical precipitation, filtration, or ultrafiltration.

Catch Basins and Sumps

A catch basin (Figure 7) is not an oil water separator or a clarifier. A catch basin is designed to keep sand and grit out of the sewer. Periodic cleanout of a catch basin is more cost effective than removing solids from the sewer at the municipal treatment plant.

A sump is designed to provide a low spot in the floor to which wastewater can drain. The water either runs directly out of it to the sewer or is pumped to the sewer. A sump usually provides no treatment.

If you are letting oil and oily water or finely divided metal-contaminated solids run from your shop floor through a catch basin or sump and into the sewer, you are probably in violation of water quality regulations. You need an oil water separator at minimum and may be forced to use chemical precipitation, ultrafiltration, or another treatment method to produce dischargable wastewater.

Filtration

Filtration uses media that let water pass through them and retain the solids that were in the water. Filters are used to remove solids from water and to concentrate dilute mixtures of solids and water by removing water from the slurries. A variety of filtration equipment is available. The size of the equipment depends on the design flow rate. For low flow rates, a filter will not take up much space in your shop. The equipment most applicable to a small shop will be described here.

Cartridge filters are relatively inexpensive and widely available. You can buy one that will handle a few gallons per minute of water for about $20 at many hardware or builder's supply stores. For about $50 at the same place, you can pick up a small pump to move the wastewater through the filter and the necessary plumbing. Larger cartridge filters are available from other suppliers.

Cartridge filters are available in different pore sizes and flow capacities. The proper pore size must be selected to effectively remove the solids. If the pore size is too large, solids can pass through the filter. If the pore size is

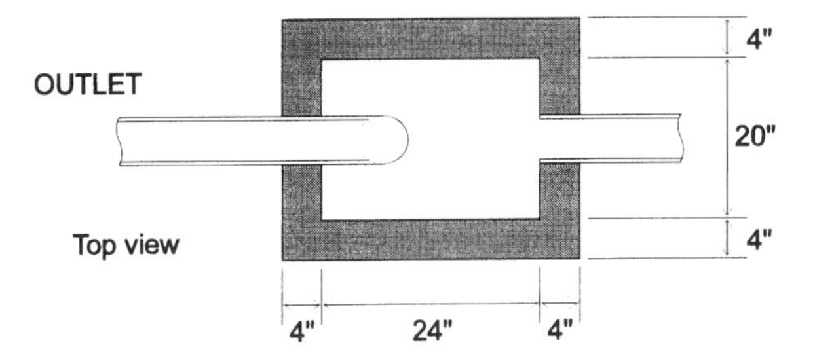

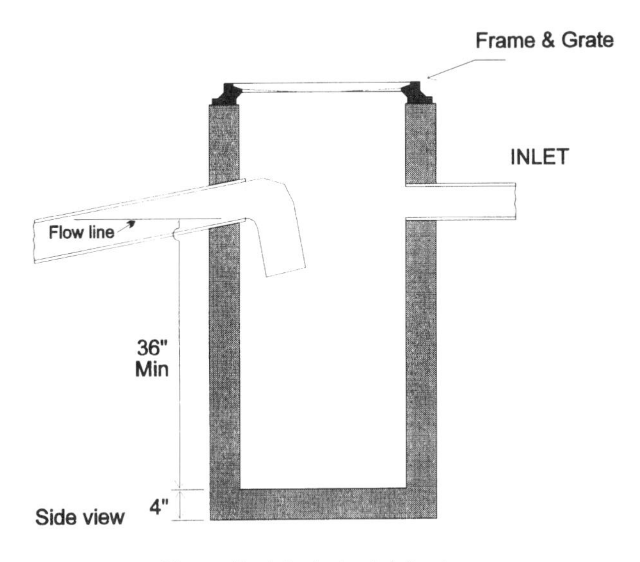

Figure 7 A typical catch basin.

smaller than necessary, water flow is impeded and the filter may need to be changed too frequently. Most cartridge filters are designed so that the water must go through the filter and cannot bypass it. One disadvantage of a cartridge filter is that it must be replaced and disposed of when loaded. Depending on what the removed contaminant is, the filter may need to be disposed of as hazardous waste.

Small sand filters which resemble welding gas bottles are available. Water is pumped through the filters at low pressure. They can handle higher flow rates than small cartridge filters and can be backflushed and reused when loaded. They are not as readily available as the cartridge water filters and cost more ($500+). The choice between a cartridge and a small sand filter depends on the wastewater flow rate and solids content.

If you are treating small volumes of wastewater you can use coffee filters, old t-shirts, or filter media from rolls to remove water from solids produced in chemical treatment. A number of filter media holders are available, or you can use a large funnel or screen.

If a filter is chosen and applied properly it can be a fail-safe system. A system can be designed so that all of the wastewater has to pass through the filter and be treated. When the filter is full, it will stop the flow of wastewater. It can then be changed out or backflushed. Some filtration systems allow the wastewater to bypass the filter when the filter becomes plugged. These systems require closer attention than a system in which unfiltered water cannot bypass.

Filtration does have its limitations. Soluble compounds, like emulsified oil or dissolved metal salts, will pass through a filter. If the pollutant that you must remove is soluble, filtration will not work unless you first treat the wastewater to make the pollutant insoluble. Small particles may also pass through a filter and will need some sort of treatment prior to filtration to make them big enough to filter. Sticky or gelatinous solids can plug a filter quickly, making filtration impractical.

Pour some of your wastewater through a coffee filter. If it looks the same going in as coming out, the particles in it are very small and may not be removed effectively by a filter. If the water comes out clear, but a sticky sludge is left on the filter, then a filter may plug quickly. If filtration looks promising, obtain a bigger filter to test out.

Chemical Precipitation

Chemical precipitation is applicable to a variety of wastewaters, including wastewater containing soluble and/or particulate metals and emulsified oil. A variety of recipes can be used. A starting recipe for chemical precipitation follows:

- Add 1 g/L ferric chloride (0.5 lb/50 gal)
- Check pH; if not <5, lower it with acid (for most consistent results)
- Add lime to pH 8.5 (at least 0.5 lb/50 gal)
- Add 10 to 20 mg/L anionic polymer (4 g/50 gal)
- Let the sludge settle and decant the treated water

The recipe can be changed if the wastewater character or discharge limitations change. You can process a batch of wastewater in a few hours and therefore increase the amount of water treated, if your wastewater flow increases.

You can do it in a 5-gal bucket, a 30-gal trash can, a 55-gal drum, a 100, and a 50,000 gal or larger tank. It could also be done in a continuous system. However, batch treatment is more practical for the typical small volumes of wastewater generated by a small shop. The wastewater must be mixed for treatment with a 2 × 4" board or tank mixer. The cost of the treatment tank and mixer depends on the size needed. Costs range from about $25 for a drum and a 2 × 4 to about $5,000–10,000 for a shop-built 1,000-gal skid-mounted system complete with air-operated pumps and mixers to $25,000 for a 10,000-gal tank with mixer and pumps. You can economize by buying used equipment. The cost of treatment chemicals for process rinsewater is minor, typically ranging from $0.005 to 0.02/gal.

Treatment of process solutions can require greater amounts of chemicals, depending on the nature of the spent solution. For example, the process chemicals in spent caustic or acid tanks must be neutralized with an equivalent amount of acid or base, and the metal salts and other contaminants must be precipitated. It can take more acid or base to bring the solution to a pH of 8 and precipitate the solids than it takes to neutralize the acid or base in the solution. The only way to tell how much acid or base you will need is to conduct bench testing.

Batch chemical precipitation does not take much time after you become familiar with the procedure. Estimate 1 to 2 hours of labor per batch to mix and add chemicals and to decant the treated water. The actual treatment time may be longer, but you do not need to watch it the entire time. If you have to bench test every batch of wastewater before treatment, you will probably spend 1 to 2 more hours per batch.

Chemical precipitation is one of the most widely used industrial wastewater treatment methods. Its primary disadvantage is that additional solids are generated by the addition of chemicals. However, the amount of solids added is about 1 g/L or 1 lb/100 gal. In many cases more solids are removed from the wastewater than are added to it during treatment.

The treatment of typical process rinsewater does not result in a lot of sludge. For example, in a 55-gal drum, about 1 in. or about 1 to 2 gal of sludge are produced. The wastewater volume is reduced about 50 times. The water can be decanted off and the sludge can be filtered and dried to remove water and further reduce the volume.

Process solutions can be treated chemically, but in many cases a large volume of sludge is produced. In some cases, as with spent caustic or acid solutions, the sludge volume can be the same or greater than the original spent solution volume. You can do bench tests to get an idea of how much sludge you will produce.

Learn to treat your rinsewater and dilute process wastewater before you commit to treating your process solutions. It is usually harder to treat process solutions than rinsewater, and the potential cost savings may not be as large. Treatment, storage, and disposal facilities build sludge handling charges into waste treatment charges, so it may not benefit you to treat solutions that produce a lot of sludge. It may cost as much to have the sludge hauled away as it would have for the untreated solution itself.

Wastewater treatment chemicals include polymers and inorganic chemicals. Polymers are chains of organic chemicals containing large numbers of charged groups. The polymers may be cationic (positively charged), anionic (negatively charged), or nonionic (no charge). Cationic polymers are typically used as primary coagulants. They are added to the wastewater to cause the small particles of oil or dirt to become unstable and stick together. Cationics are typically most effective at neutral or slightly acid pH. Anionic polymers are typically used following primary coagulant addition to cause the particles to form even bigger particles which settle rapidly. Anionics are typically most effective in basic solutions. Nonionic polymers are not as commonly used for wastewater treatment as the cationic and anionic polymers. Many different types of polymers are available. Polymer suppliers can supply samples for you to evaluate.

Polymers can be used alone to cause oil or solids to coagulate. However, polymers are very sensitive to wastewater chemistry and may not produce consistent results because wastewater characteristics tend to change. Oily wastewater may have different quantities of detergents from batch to batch, and different polymer doses or polymer types may be required to treat each batch. Do not rely on testing one batch of wastewater to determine if a polymer works. If possible, you want to install a system that you do not need to adjust for every batch of wastewater. Polymers alone are not as reliable as when used with inorganic coagulants unless you have a very consistent wastewater.

The most effective broadly applicable and consistent chemical treatment relies on an inorganic coagulant addition (with acid addition, if needed) to drop the pH followed by the addition of lime [$Ca(OH)_2$] or caustic (NaOH) and the addition of a high molecular weight anionic polymer. The more commonly used inorganic coagulants include the metal salts ferric chloride ($FeCl_3$), ferrous sulfate ($FeSO_4$), and aluminum chloride ($AlCl_3$).

All of these chemicals are readily available. Ferric chloride can be bought by the gallon in 35% solutions for about $10/gal or dry by the pound. Aluminum chloride and ferrous sulfate also be obtained dry by the pound. A 50-lb sack of ferrous sulfate runs about $20. A 50-lb sack of hydrated lime (agricultural grade) can be bought at a cement yard for about $6. Caustic is available in 30 or 50% solutions by the gallon or drum and dry by the pound. Take special care when handling caustic and protect your operators from dust from lime and other dry chemicals.

The coagulant and dose that works best depends on the wastewater. The dose that works most of the time is 0.5 g/L as $FeCl_3$ or about 1 cup of 35% $FeCl_3$ solution per 50 gal. You can conduct bench tests to optimize the $FeCl_3$ dose.

Treatment with inorganic coagulants and lime or caustic followed by a polymer causes a complex series of reactions which effectively remove emulsified oils and soluble and particulate metals. When the metal salts are added to the wastewater, the pH falls; if pH does not fall below 5, then acid should be added to the solution to bring the pH down. If you know you will need to add acid, then add it before the salts. Oily wastewater will not break as cleanly, and maybe not at all, if the pH is not dropped into the acid range. Some colloidal solutions such as paint booth wash water will not respond. You need to go down and come back up. The acid pH causes the particles in the water, which are negatively charged in alkaline water, to become neutralized. The neutralized particles do not repel each other as strongly and usually start to collect into larger particles. When you add the metal salts and drop the pH, you are likely to see particles grow in the wastewater. This is the first step in removing them from the water.

The amount of acid that you add to get the pH of caustic wastewater to drop below 5 depends on the wastewater. It can range from 0 to 1 oz/gal (or more) of concentrated acid, depending on how much base is in the wastewater. The only way to tell how much acid must be added is to add acid to a sample of wastewater and find out. You cannot predict the amount of acid needed from the initial pH.

After a few minutes to allow the reactions to occur, add lime or caustic to bring the pH up to about 8.5. The amount of lime or caustic that must be added to bring the pH up to 8.5 to 9 cannot be predicted and must be determined by testing.

The final pH is very critical, and if it is much below 8, metals will not precipitate effectively. Lime tends to form a more stable precipitate than caustic. When lime is used the process is less sensitive to overshooting the pH than it is with caustic. Use lime if possible, but take care to protect yourself and others from the dust. (Wear gloves when handling it, otherwise your skin will dry and crack.)

At least three things happen when the pH is brought back up. Particles that were positively charged at the acid pH become neutralized, and again the neutral particles tend to stick together. The metals in the water form hydroxides and become less soluble. The metals precipitate and tend to collect small particles in the water. The iron (ferric or ferrous) or the aluminum that was added forms fresh hydroxide, which effectively adsorbs soluble metals from the solution.

Iron and aluminum hydroxide-coated sand, activated carbon, or alumina can be used in columns like ion exchange resins to adsorb metals. The columns can be used to remove soluble metals but not particulate. In fact, the water must be treated ahead of time to remove solids. If you must use chemical treatment to remove finely divided solids, then you have no need for an adsorption column. You are, in effect, creating the adsorbent in the solution instead of using a column.

After the pH has stabilized between 8.5 and 9.5, the mixing can be stopped and the solids allowed to settle. If particles are not visible in the water, the process did not work. The particles that settled should be visible, leaving clear water above.

The addition of a small amount of anionic polymer, typically 10 to 20 mg/L, after the pH has been brought up, can greatly aid clarification of the water. The polymer attaches to particles and connects them, causing the formation of larger particles. The larger particles sweep up other, smaller particles. Polymer addition both speeds up settling and improves solids removal.

Ferric chloride and lime followed by an anionic polymer can treat a wide variety of wastewater including oily water, metal-bearing water, food processing water, and arsenic-contaminated groundwater. Chemical precipitation will remove emulsified oils, metals, and particulate from wastewater. It is flexible and can be accomplished with simple equipment.

Evaporation

Evaporation is simply boiling off the wastewater, leaving the contaminants behind, and reducing the waste volume. It can be used for oily or metal-contaminated wastewater. It is not the best option for the treatment of wastewater that contains organic contaminants, such as solvents, that will boil off with the water. If the waste contains oil or other nonvolatile organics, you should avoid evaporating it to dryness. If you boil it down too far it can overheat and release smoke and fumes. Air pollution control agencies are not usually concerned with the discharge of water vapor unless it is creating a nuisance problem, but they are concerned with the emission of air contaminants.

A number of different types of evaporators are available. The simplest commercial evaporators consist of a tank heated from below with gas burners (or electric elements), having a vent and blower to exhaust the water vapor outside the shop. As the wastewater evaporates and contaminants are concentrated, scale or adherent solids tend to form on the sides and bottom of the tank. The coating insulates the tank and hinders the flow of heat into the tank, making the evaporation less efficient. Some evaporators have devices which attempt to reduce the detrimental effect of the solids coating. Oil may float to the surface of the wastewater and hinder evaporation. Removal of the floating oil may improve the efficiency of the evaporator.

How well evaporation works for you depends on the nature of your wastewater. Investigate evaporators being used for wastewater similar to yours before you buy one. They are used by machine shops to deal with rinsewater from their caustic hot tanks and jet washers. Metal finishers use evaporators to concentrate rinsewater from certain process tanks to return it to the process tanks, recycling the process chemicals which were dragged out in the rinsewater. Evaporation may be used to further concentrate the residue produced by another treatment process. For example, ion exchange regenerant water or chemical precipitation slurry may be evaporated.

You can boil off a few gallons per day inexpensively with a homemade boiling pot. You could use a cooking pot on a stove. You can spend $2000 to $5000 for evaporators with 50 to 100 gal per shift capacity. Or you can spend tens of thousands of dollars for units in the 1000+ gal per shift range. The major operating cost is producing the heat needed for evaporation. The major maintenance costs include periodic replacement of heating coils and the evaporator bottom in some units. One of the advantages of evaporation is that it may not take much attention to run. The biggest chore, if evaporation is right for your situation, is removing the concentrate.

Oxidation

Chemical oxidation can be used to destroy certain organics such as phenol and inorganics such as cyanide. Common oxidants include chlorine bleach, hydrogen peroxide, potassium permanganate, and ozone. Chlorine is not a good choice for phenol oxidation because chlorophenols are produced. It is typically used for cyanide oxidation by metal finishers prior to metal precipitation.

Oxidation can be done in batches as in chemical precipitation, but it is more complicated. Oxidation requires careful control of reaction conditions. In the case of cyanide oxidation care must be taken to control the pH so that it does not drop below 10 until the cyanide is oxidized. Alternatively, hydrogen cyanide gas can be liberated from the solution. If you need to do oxidation, you have more homework to do than for chemical precipitation.

Reduction

Reduction is typically used to change hexavalent chromium (Cr^{6+}) to trivalent chromium (Cr^{3+}) prior to chemical precipitation. Cr^{6+} will not precipitate as a hydroxide, but Cr^{3+} will. Therefore, reducing agents such as sodium bisulfite or ferrous iron must be used. Ferrous sulfate addition at a pH of <5.5 can reduce Cr^{6+} and also provide a metal coagulant that will aid in the precipitation of regulated metals. Ferrous iron serves several purposes and can be used instead of sodium bisulfite to reduce Cr^{6+}. Reaction conditions are critical for reduction processes, and again you will have to do more homework than you would if chemical precipitation alone will work for you.

Air Stripping

Air stripping is used to blow small amounts of volatile organics, such as toluene, methylene chloride, and xylene out of the water, or in reverse, to adsorb contaminants such as ammonia from air. It is not applicable to blowing off nonvolatile compounds, such as phenol, or volatile organics, such as acetone, MEK or isopropyl alcohol, which are very soluble in water. In order to strip these types of compounds the water must be heated to 100°F or more, which is a potentially expensive step. If your wastewater contains volatile organics you should try to keep them out of it by changing your process chemicals or collecting organic contaminated wastewater separately from other wastewater.

If you cannot keep the volatile organics out of the wastewater and their concentrations exceed your discharge limitations, you may want to consider air stripping. However, your local air pollution control agency will be concerned about the amount of organics you blow into the air and may require treatment of the air stripper exhaust.

In some cases the amount of volatiles that the air stripper emits will be less than the allowable amount and you will need no additional treatment. If you must remove the stripped volatiles from the air stream, then you could investigate activated carbon canisters. Other volatile removal technologies, such as catalytic oxidation or incineration, are likely to be cost prohibitive for a small shop and are typically used at large facilities.

Air strippers are designed to provide contact between the wastewater and an air stream to allow organics to be removed by the air. Several designs are available. The most common is a packed tower with 15 to 20 ft of packing. The water is introduced at the top of the tower and runs down through the packing which spreads it out and creates a large water surface area. The air is blown in at the bottom and collects organics as it moves up through the tower. Air:water ratios of 100 to 200 cfm air:cfm water or about 10 to 20 cfm/gal of water are used. Solids and scale-forming chemicals will plug up the packing and they must be removed from the wastewater before it reaches the stripping tower. A 50 gal/min air stripping tower costs $10,000 to $20,000. Other systems use baffled trays or pipes to distribute the water. Tray systems may tend to be more tolerant of solids and scaling. A 10 gal/min tray system costs about $5,000. These systems tend to be less efficient at removing organics, but can be used if discharge limits are not very low.

Air stripping is a well-developed technology. Manufacturers and suppliers have a lot of operating data and understand the limitations of the systems. Contact suppliers for information if you need an air stripper.

Activated Carbon Adsorption

Activated carbon adsorption is also a well-developed technology. It is widely used to remove organics from wastewater, and can be used to remove organics such as phenol, toluene, trichloroethylene, and naphthalene. It is much less effective at removing water-soluble organics such as acetone, MEK and methylene chloride.

The absorption isotherm of a compound is a measure of how much of an affinity it has for activated carbon. The isotherm depends on the nature of the compound as well as on the type of the activated carbon. Isotherms are available in the literature for a number of organics.

A number of suppliers offer easy to install activated carbon units. The units can be shipped back to the suppliers when depleted for regeneration and reuse for the cost of shipping and about $2/lb of carbon. A drum-sized unit, holding about 100 lb of activated carbon, which can treat about 10 gal/min costs about $500. A 4 × 4 skid-mounted transportable unit, holding about 1000 lb of activated carbon, for flow rates of about 50 gal/min, costs about $4000.

The spent activated carbon may be designated as hazardous waste depending on what it removed from the wastewater. You must legally dispose of the spent carbon if it is not reclaimed by the supplier. Include the disposal cost in your treatment system evaluation.

The time between carbon change out depends on the type and amount of organics in the wastewater. The suppliers can estimate the useful life of a unit based on the wastewater characteristics. Solids will filter out in the units, reducing their capacity or plugging them up. Therefore, solids must be removed before the carbon units.

Ion Exchange and Adsorption by Other Media

Ion exchange resins are typically used to treat wastewater which contains soluble metals and is relatively free of particles. Ion exchange is well suited for the treatment of rinsewater from certain metal-finishing processes such as copper-, gold-, or silverplating. The adsorbed metals can be removed from the resin by regeneration and potentially recovered for reuse. In effect ion exchange concentrates the metals in the wastewater.

It does not work as well when rinsewater streams are mixed. Chemicals for various processes can combine to make the wastewater unsuitable for ion exchange. For example, pH is critical and rinsewater from mixed processes can vary widely. Also, chelating and complexing agents can interfere with ion exchange, and the system must be designed to accommodate them. If you are treating a mixed stream the character of the wastewater will be less predictable and ion exchange may not be as successful. It works best when dedicated to a particular process rinse tank. Installing a separate unit for each metal-bearing rinse tank increases the initial cost, but results in a much more reliable system than if one unit is used for all rinse tanks.

Ion exchange resins are susceptible to fouling by organics such as oil and detergents and certain inorganics such as iron. The contaminants cannot be removed easily from the resin and shorten its useful life. The wastewater must be relatively free of these contaminants for ion exchange to be viable.

A variety of adsorption media function much like ion exchange resins. These include iron hydroxide-coated sand or alumina, peat moss, biological materials, and proprietary materials. These media can be less susceptible to fouling than ion exchange resins. Because this is an emerging technology these media are not as well established in the marketplace as ion exchange resins and not as much information about them is available. They have been used in specialized applications.

Ion exchange and adsorption columns are designed to remove soluble metals, not solids. In fact, solids will coat and obstruct the media. The water must be treated to remove solids before ion exchange or adsorption. In some cases, after this pretreatment, the water will meet discharge limits and will not need further treatment.

Resin and adsorbent selection is critical, and the wastewater conditions must be carefully controlled. Contact system suppliers to determine if your wastewater is amenable to ion exchange or adsorption treatment and what resin to choose.

Electrolytic Recovery

Under certain conditions metals can be removed or plated out of solution using an electrolytic process. Metal finishers use specially designed electrolytic cells, also known as electrowinning cells, to remove metals from rinsewater. The cells oxidize the metals onto a cathode. Other types of cells use electrodes and membranes to pull metals such as chromium (Cr^{6+}) through a membrane to concentrate it for recovery.

Electrolysis takes time. The unit must be designed to allow the reaction to proceed to the extent desired. To electrowin metals from dilute rinsewater a large electrode surface area is required because of the low efficiency of the process. Electroplating electrolytic recovery units are typically used in conjunction with recirculating rinse tanks to provide for long contact time.

Electrolysis works best when the water does not contain particles, oil, or biological material. Particles, oil, and biological slime can coat and foul the electrodes in the cell, causing the efficiency to drop. Slime that adheres to the plates can slough off and cause discharge limits to be exceeded.

Membrane Separation

The different membrane technologies include ultrafiltration, reverse osmosis, ion exchange, and electrodialysis.

Ultrafiltration can be used to remove water from wastewater containing emulsified oil, reducing its volume. The water passes through the membrane and oil particles are retained. The water that goes through the membrane can meet FOG discharge limitations. In some cases, depending on the nature of the contaminants, metal particles also may be excluded from the membrane and removed from the water passing through it. However, soluble metals pass through the membrane. For example, ultrafiltration is not applicable for the treatment of copper-plating rinsewater, and water-soluble machine coolants can pick up zinc during use and the zinc may get through the membrane. Soluble organics such as phenol, toluene, acetone, and MEK also pass through the membrane.

The ultrafiltration membranes need to be cleaned and backflushed regularly to operate efficiently. Over time they slowly plug up and periodically need to be replaced. Some wastewater contains particles that are the same size as the membrane pores. For example, automotive antifreeze contains silicates which apparently fit into the pores. The particles will quickly irreversibly plug the membrane. In this situation ultrafiltration is not feasible.

A 1-gal/min ultrafiltration unit costs about $2,000. A 10-gal/min system costs about $15,000. The 1-gal/min system is designed to fit on a 55-gal drum. The 10-gal/min system is skid mounted and takes about 2×4 ft of floor space and is about 5 ft high.

Reverse osmosis, ion exchange, and electrodialysis may be applied in small shops, especially those that have plating processes, but are typically used by larger facilities. Reverse osmosis removes soluble compounds such as sodium chloride. A concentrated reject stream is produced and must be disposed of. Wastewater must be treated to remove solids prior to reverse osmosis or the membrane will plug quickly. Ion exchange and electrodialysis separate soluble charged ions from solution by imposing a concentration or electrical gradient. Solids can also interfere with these processes.

Thermal Treatment

Thermal treatment processes destroy the waste by burning it or exposing it to high temperatures to break it down. Some wastes can be burned for energy recovery in industrial furnaces, but a small business is not likely to have a suitable industrial furnace, or cement kiln, on-site. Hazardous waste incinerators, another means of thermal treatment, are very expensive to install, permit, and operate, and are beyond the means of a small business and typically not cost effective, unless a facility generates a large amount of waste.

Thermal treatment by an off-site facility may be feasible for some treatment residue but it is among the last choices for the treatment of wastewater. Some processes, such as wet air oxidation, are used for difficult-to-treat wastewater, but the equipment is expensive. If you need to use thermal treatment for your wastewater it will probably be most cost effective for you to send it off-site.

Biological Treatment

Biological treatment is used to remove organic compounds from wastewater. It is most applicable for wastewater that contains a relatively constant source of biochemical oxygen demand (BOD) and very low concentrations (on the order of 1 mg/L) of toxic metals. A surge tank to equalize wastewater flow and concentration variations can help the treatment system work effectively.

Municipal treatment plants commonly use biological treatment to treat domestic sewage. Biological treatment is cost effective for some industrial wastewater, including some food processing, pulp and paper, and chemical manufacturing wastewater.

Biological treatment has limited applications in small businesses. It costs about $5,000 for a 10-gal/min system and $25,000 for a 100-gal/min system. The size of the system is dependent on the water flow rate, and the type and amount of BOD. The microbes require a relatively constant source of wastewater and BOD. The wastewater from small shops tends to be generated sporadically, and may contain metals and oil that are detrimental to the microbes.

A food processor may consider biological treatment. However, municipal plants are designed to treat BOD and they handle large volumes of wastewater. Many municipalities will accept BOD-containing wastewater and charge the discharger for its treatment. For a small company this is usually more cost effective than in-house treatment. If you are a food processor and do not have the advantage of discharging to a municipal treatment plant, look into buying a packaged biological treatment system or building a larger system, if needed.

SYSTEM DESIGN OPTIONS

Do-It-Yourself System

You can design, build, and install a wastewater treatment system yourself. This is appropriate if you have a well-characterized wastewater which can be treated using gravity separation, filtration, or chemical treatment. You can assemble the system from off-the-shelf components. Use this book as a guide. If you do it yourself you will presumably understand how it works and what it does.

Regulatory agencies tend to be receptive to the initiative of a pollution control manager or business owner. If you do a good job of developing your system and document the design process adequately, chances are you can obtain approval to install it yourself. You must provide enough information to convince the regulators that you know what you are doing.

Purchased System

You can have a vendor or consultant come in and do it all for you. They will characterize your wastewater and recommend what type of system to install. However, you still must evaluate their proposal. You may end up spending more for your system than if you developed it yourself because you obviously must pay for their overhead. However, if you pick the right supplier you benefit from his/her experience and you could end up with a better system than you could put in yourself.

If you need specialized equipment such as an ultrafilter, air stripper, or carbon adsorption unit, you are probably better off buying an off-the-shelf system than trying to assemble one yourself. However, you can evaluate the feasibility of the equipment before you buy it and fit a system together that works for your shop.

Haul It Away

You should determine the cost of having a licensed facility haul and treat your wastewater. This will give you a basis for evaluating the cost of your treatment options as you develop them. It may even help make your decision easier to live with by providing perspective. It is not realistic to compare treating the wastewater with the cost of letting the wastewater run down the sewer untreated.

INFORMATION SOURCES

Contact Shops

Other shop managers have thought about wastewater treatment. Many of them are happy to share information about environmental issues. Call around to other shops that have processes similar to yours. Find out who has installed a treatment system and ask if you can come see it. They are proud of what they have done and are usually willing to show you. Find out what technologies are in use in shops like yours. Put them on your list of alternatives.

Contact Trade Associations

Your trade association exists to help enhance the competitiveness of your shop. Many trade associations have become aware that it is in the best interest of their members to address public concerns and environmental regulations. They have realized that it is good public relations to identify and fix problems, and to make known their conscious efforts on behalf of the environment.

Your trade association may be able to refer you to other shops that already have installed wastewater treatment systems. It may hold environmental workshops in conjunction with regular meetings, it may have information about equipment suppliers, and it may be involved with regulatory agencies helping to interpret the regulations as applied to your trade.

Call your trade association. If you are not a member of a trade association, ask other shop managers if they are. Find out if a proactive association exists that you can get involved in. You could save time by getting a head start on your problems. At least you can find others with whom to commiserate.

Contact Suppliers

Many suppliers offer a variety of treatment equipment. They will be glad to tell you what they have. Call and ask for information. Include it in your list of alternatives. Working with suppliers to obtain information and equipment can be challenging. (See Chapter 8.)

Go to the Library and Bookstores

Reference books, textbooks, and Environmental Protection Agency (EPA) manuals suggest an assortment of wastewater treatment methods. Probably no one book will answer all of your questions, or will use terms that you can understand and apply to your situation. Visit your local library or bookstore and see what they have. Ask other business owners and plant managers what books they have found to be useful. See the book listing in Chapter 9. A number of publishers will give you 15 to 30 days to review a book to determine if it meets your needs.

Contact Regulatory Agencies for Information

Regulators specialize in ensuring that businesses comply with the rules. They get around to a lot of businesses and see environmental problems and what the businesses have done to deal with them and stay in compliance. They also see what does not work well. Ask your pretreatment authority if they have information that can help you. They may put you in touch with a shop manager with similar problems, so you can talk with a peer, or they may be able to give you a list of vendors and consultants.

Some agencies have a technical assistance team that will come to your shop and give you advice without the threat of a penalty notice. In some cases they have developed best management practices (BMPs) for your wastewater, which tell you what you need to do in order to discharge the wastewater to their collection system. If this is the case your alternatives are narrowed considerably and you can focus your efforts toward complying with the BMPs. They may have developed wastewater treatment guidance documents which examine treatment alternatives for shops similar to yours.

LIST THE ALTERNATIVES

List the alternatives that you have identified to document your research. The list will be included in your treatment plant design report. This will help to show the basis of your eventual selection. Include the alternatives that are not feasible. These alternatives will help to put the others into perspective. The reasons something will not work are just as important as why something will work. Knowing what the problems are ensures that the problems are addressed and resolved. Bad alternatives will stand out and the reasons that they do will be apparent. The best choices also should stand out.

5 EVALUATING THE ALTERNATIVES

INTRODUCTION

This chapter explains how to go about evaluating the wastewater management alternatives that you have identified. Essentially, you need to develop objective information to compare the alternatives and to decide which one is best for your shop. Many alternatives will work. Each has its own set of advantages and disadvantages. You need to find out what they are. In the final analysis your choice will depend on what makes sense to you. The objective is to understand your situation well enough to make a good judgment. When evaluating alternatives, do not rely on hearsay; investigate it yourself. Steps associated with evaluating alternatives are presented in Table 17.

EVALUATION CRITERIA FOR A WASTEWATER TREATMENT SYSTEM

Define the criteria you will use to evaluate the alternatives, including:

Effectiveness and economics
Reliability of the system
Flexibility of the system
Capital and operating costs
Space requirements
Operating and maintenance requirements
Disposition of waste treatment residual

Of each system under consideration ask: Does it meet discharge limits consistently? Can it adapt to changing workflows and waste characteristics? How much does it cost to install, run, and maintain? Will it fit in the available space? How much training and skill does it take to run? Can you and your operators understand the system well enough to make it work? What kind of treatment residue does it produce and how will it be disposed of?

Table 17 Steps Associated with Evaluating Wastewater Management Alternatives

Define the criteria you will use to evaluate the alternatives
Evaluate alternatives to a pretreatment system
Evaluate potential pretreatment technologies
• See it in action
• Do a bench test
• Conduct a pilot test

How you prioritize the various criteria will depend on your circumstances. No standard treatment system will work in all situations. Depending on who you talk to, you will get a variety of opinions about a given system or technology. That makes the selection of a system seem intimidating. However, with a systematic approach you can sort out and evaluate alternatives.

As you get into the process and learn what is important to know, you will formulate good questions. The most important things to consider are covered in this chapter. Ensure that, at a minimum, you address them. Above all else, the system must work. The last thing you want to happen is that the treatment system that you spend $10,000 or $1 million for, does not produce wastewater that meets discharge limitations, or that it meets the limits only if you spend all of your time making it work, or you make a process change and it cannot handle the new wastewater. The purpose of evaluating alternatives is to improve the chances that you make a good investment and limit your risk. The importance of doing your homework cannot be overemphasized.

Various waste management options are described generally and their advantages and disadvantages are explained in the context of a shop's constraints.

EVALUATION OF ALTERNATIVES TO A PRETREATMENT SYSTEM

Process Changes

If you change your process to eliminate out-of-compliance wastewater then you will not need to deal with the wastewater. However, you may end up generating other waste or using a process chemical that presents greater or unknown risk to you and your co-workers' health and safety.

For example, as described in Chapter 4, the trend is to replace common solvents such as methyl ethyl ketone (MEK), lacquer thinner, and methylene chloride (dichlorobenzene) with solvents that are not as strictly regulated. The health and environmental impacts of those common solvents have been extensively evaluated and regulatory agencies have developed rules for their use. Many of the alternative solvents have not been used as widely and have not been evaluated as thoroughly. A variety of glycol ethers, known generically as cellosolves, are finding their way into cleaning formulations. Not as much information about their environmental impact is available as for some of the more common solvents. Therefore, it may be hard to assess the risks of alternatives to your co-workers and to the sewage collection and treatment system. This is a factor that you need to consider when changing the solvents you use.

Citrus-based cleaners contain d-limonene or a similar derivative. d-Limonene is related to α-pinene, a component of turpentine and has similar physical properties. It is not soluble in water and has a flash point of about 130°F. It is an irritant to skin and mucous membranes (the nose and lungs). The long-term health effects of d-limonene are unknown. Water-soluble formulations with d-limonene contain surfactants that disperse the d-limonene in the water much like detergents cause oil to be rendered mixable with water. The d-limonene may be released from the formulation in use or in the sewer system and present worker health and safety hazards. Some shops that have tried citrus-based cleaners have stopped using them because workers did not like the smell. Citrus-based cleaners may work for you, but you need to be aware of the disadvantages.

The process change may also cause the workflow to change. For example, a shop replaced a solvent-borne adhesive with a water-borne adhesive. The solvent-borne adhesive was cured in an oven in about 15 min. The water-borne adhesive took 1 or more hours to cure, resulting in an unacceptably low production level. Solvent use was reduced, and equipment cleanup produced sewerable wastewater, but the shop could not be competitive under the circumstances.

It some cases you may change a process to eliminate the generation of an air pollutant and end up generating a water pollutant. This is the case if you change from solvent degreasing to aqueous cleaning. Now you have wastewater to worry about instead of spent solvent and have to weigh the alternatives.

The boatyard and shipyard industry provides an example. To refurbish a hull, the old coating is removed and the hull is repainted. In the past media or sandblasting was commonly used to remove the coating. The dust produced was hard to control. Time-consuming steps had to be taken to set up enclosures to capture the dust. A small hole in the enclosure could lead to fines for noncompliance. Pressurized water blasting is the current preferred method of removing the old coating. Any particles produced are captured by the water spray and dust generation is minimized, but wastewater is produced which must be treated prior to discharge. In this case handling the wastewater is cheaper and easier than trying to contain sandblasting dust.

Process changes are made for a variety of reasons. Sometimes a process change is mandated by regulations and you have no choice. Sometimes you change a process to eliminate a regulated chemical and reduce your environmental liability, or you may change a process to make your waste easier to manage.

You need to consider the impact of a process change on your shop operations, waste streams, and proposed treatment system. Before you install your treatment system, think about what changes you may make and how they will change the character of your waste. The changes may make the treatment system unnecessary or ineffective. Your wastewater treatment system should be developed in the context of the larger environmental compliance plan for your shop. The treatment system is only one piece of the plan.

Chemical Recovery

Wastewater includes rinsewater, spent process solutions, and fluids, such as automotive antifreeze, which are removed from equipment being repaired or serviced. Recovery and reuse of process chemicals follow the elimination of undesirable chemicals in the pollution prevention hierarchy. A variety of recycling systems are available for various waste streams.

The feasibility of recovery and reuse depends on how sensitive the process is to contaminants and variations in solution composition, how hard it is to install and operate the recovery equipment, the cost of new chemicals, and the cost of disposal of spent chemicals. Also, when you recover chemicals or wastewater for reuse, you may generate another waste stream that you will need to manage.

The same methods used to treat wastewater for discharge may be used to recover wastewater and chemicals for reuse. The major difference in the evaluation is that the shop's process water quality criteria are used in place of the sewer discharge requirements for wastewater recovery, and process specification requirements are used for process solutions, fluids, and chemicals recovery. Therefore, recovery methods are not called out separately in this section, but are implicitly included. Just remember that some of the treatment methods also may be used as recovery methods.

EVALUATION OF ALTERNATIVES

See It in Action

One of the best ways to evaluate a treatment technology is to see it working in a shop similar to yours. Formulate the questions that you need answered before you visit the shop. Find out how much the system cost and how long it took to install. How much wastewater do they treat each day? Is the wastewater similar to yours? How long has the system been in place and what problems have been encountered? What shop and process modifications were needed to make the system work? Find out the background of the treatment system operator and how much training was needed to learn to run the system. Determine how much time and attention it takes to run the system. Ask the manager what level of service the vendor supplied and if he/she is satisfied with the vendor's support. Also, find out why the manager thinks the system is working as it should. What kind of test data are available? Has the pretreatment authority taken a look at the facility and found it satisfactory?

Take a look at the system and determine if you could handle it in your shop. Can you understand the principles of operation well enough? Are you willing to handle the chemicals needed and the waste residue produced? Are the purchase, operating, and maintenance costs within your budget?

Do a Bench Test

Bench testing is a good way for you to get a feel for the treatment technology and see if it is applicable to your wastewater. A bench test is a small-scale simulation of the treatment process. You can bench test any process. However, the equipment you need depends on the type of treatment technology.

The purpose of a bench test is to evaluate the technology under controlled conditions. You can test portions of the same wastewater sample while changing treatment parameters. For example, you can run tests to compare the effect of different chemicals and their doses on treated water quality. You can assess the effect of pH on final metal concentrations. You can estimate the amount of acid or base needed for treatment. You can see if filtration is applicable to your wastewater or if a filter lets particles through or quickly plugs. You can borrow a small ultrafilter to see if it can produce effluent of acceptable quality.

Bench testing is a good first step in learning about a treatment technology. It can be done at relatively low cost and personal experience in how the technology works is gained. On the bench you can more easily assess the effect of treatment method changes than in a larger treatment unit. What you learn will help you to troubleshoot the full-scale system. Ask your vendor how to bench test the system and conduct a bench test of your own.

Chemical Precipitation Bench Test Example

The following is an outline of how to go about conducting a chemical precipitation bench test. It tells you what equipment you need, how to use it, and what to look for during the test.

Equipment Needed

An ideal bench testing setup is shown in Figures 8 and 9 and components are listed in Table 18 with price ranges. It consists of the following equipment: a set of 500 ml glass beakers, a pH meter, a magnetic stirring apparatus, magnetic stirring bars, some plastic pipettes, a balance, the test chemicals, bottles or jars to dilute them in, and representative wastewater samples. You can purchase the test equipment from a laboratory supplier for as low as $300 if you shop carefully. Check at surplus sales for used equipment. The author found the triple beam balance shown in Figure 8 for $20 at a garage sale. You can purchase test chemicals, in small amounts, from the same supplier or potentially more cheaply, in larger quantities, from a chemical supplier.

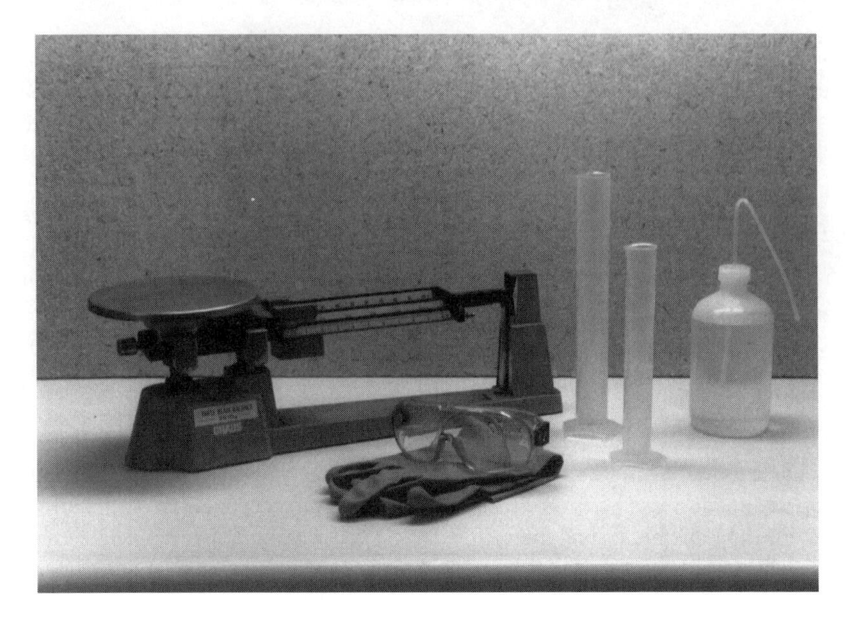

Figure 8 Bench test equipment used to measure the amounts of treatment chemicals added.

Figure 9 Bench test equipment used to mix the test solutions and determine their pH.

You can do the testing in jars (bench testing is also known as jar testing) and mix the solutions with a spoon. You can use pH paper instead of a pH meter and a measuring cup to determine wastewater volumes. However, you will still need some way to weigh out your treatment chemicals.

Table 18 Bench Testing Components with Price Ranges

Bench Test Component	Price Range
• 500 ml glass beakers	$22/6
• Graduated cylinders (10, 25, 100, and 250 ml)	$5–10 each
• pH meter	
Pocket probe type	$30–200
Portable meter	$200–900
Benchtop meter	$700–2000+
• Magnetic stirring apparatus	$100–200+
• Magnetic stirring bars	$5–10 each
• Plastic pipettes	$25/500
• Balance	
Triple beam	$100–200
Electronic	$700–1700+
• Test chemicals	Varies
• Sample and chemical bottles	
1 L plastic	$25–35/12
1 L glass	$30–60/12
250 ml plastic	$12–15/12
250 ml glass	$25–40/12
25 ml VOA bottle	$100–150/100
• Representative wastewater samples	The time to get them
Budget Component	**Price Range**
• Jars	Free
• Spoon	Free
• pH paper	$5–15/100 tests
• Measuring cup	$1–5
• Diet scale	$15–30
• Sample bottle	Free from test lab

Ferric chloride and lime (calcium hydroxide) work well on a wide variety of wastewaters. Therefore, you should plan to try them. Aluminum chloride or ferrous sulfate work better in some applications and their effectiveness could be compared to the ferric chloride. Sodium hydroxide (NaOH), also known as caustic, can be used in place of the lime, but lime tends to work better in many instances and provides more consistent results. These chemicals are commonly available. You can buy all of them in dry form. Sodium hydroxide and ferric chloride are also readily obtained in solution.

Magnesium hydroxide may be used in place of lime or caustic, but it costs more and is less readily available. In some cases it produces less sludge than lime or caustic. However, the volume reduction may not be significant because solids already present in the wastewater precipitate adding to the sludge, and you probably will not generate much sludge.

You will not need more than 0.5 lb of each chemical to get started. If you visit shops with chemical precipitation systems you may be able to get some of their chemicals for your testing. Remember you will have to figure out how to dispose of any chemicals you do not use. Until you decide on what chemicals you will use, do not buy large quantities of them. For example, do not buy 5 gal of 50% caustic or 50 lb of lime unless you know someone else who can use it if you do not use all of it.

Call some treatment polymer suppliers and get some high molecular weight anionic polymer. They may provide free samples or be willing to sell you a small amount. You will only need a few ounces of polymer for bench testing.

Preparing for the Test

Obtain the test equipment and chemicals. Calibrate your pH meter, if you have one, according to the instructions. Prepare the chemicals. Get your wastewater sample.

The best way to add chemicals is to mix them in tap water first in known concentrations. The premixed chemicals dissolve more easily in the wastewater than the dry chemicals and it can be more convenient to measure out the treatment chemicals by volume than weighing them out for each test. Use pieces of aluminum foil or plastic for weighing dishes. Moist ferric chloride will attack aluminum. If your scale will not weigh out in tenths of grams then you must make up solutions because you will not be able to accurately measure out the small amounts needed. A postal or diet scale can be used to weigh out ounces of chemicals.

You can work with pounds and ounces but I prefer working with grams and milliliters. Once you understand the metric system the calculations are easier. To help you make the transition to the metric system, if you are unfamiliar with it, the necessary calculations are shown using both systems. It may become apparent to you that the metric system is actually easier to use for bench and pilot testing work. You can convert the results to pounds per gallon doses for use in full-scale treatment.

Make up separate 5% solutions of each treatment chemical that you will use, ferric chloride, aluminum chloride, ferrous sulfate, lime, or sodium hydroxide. Add 6 oz to 1 gal or 50 g to 1 L of water. This is equivalent to about 20 lb/50 gal. Each 10 ml (one third of an ounce or 2 tsp) of solution will then contain 0.5 g of the chemical. Shaking or stirring the solutions will aid dissolution of the chemicals. The caustic will take a while to dissolve and the lime will not quite dissolve, but will be a slurry. If you must add acid to your wastewater to drop the pH during treatment, prepare a 5% acid solution.

Anionic polymer may be provided as a dry powder or as a solution. In your bench testing you can add it as received to see if it will work, but it is more advantageous to dilute it to 0.1% (1 g/L or 0.13 oz/gal) to allow it to unwind in solution before you add it to the wastewater. Each 10 ml of solution will then contain 10 mg of polymer. Use warm water to dissolve the polymer. It will take a while to disperse in the solution and the solution will become viscous like syrup.

Run a Blank

To become acquainted with chemical precipitation first run through the method using clean tap water. Try the following recipe:

- Place 500 ml or 16 oz of tap water into a beaker or jar
- Test the pH; it should read about 6 to 7

- Add 10 ml or 2 tsp of 5% ferric chloride solution, a dose of about 0.5 g/L or 2 g/gal
- Mix the water; it will be light brown in color
- Test the pH; it should be <5
- Add lime slurry until the pH is about 8.5; it should take about 10 ml or 2 tsp; the lime reaction is slow, so take your time and try not to overshoot the pH
- As the pH rises, the ferric iron combines with the hydroxide being added and precipitates as small particles of iron hydroxide the color of rust
- After you have adjusted the pH to about 8.5, stop mixing the solution; the particles should slowly settle
- After you observe the settling characteristics, gently mix the solution and add 20 ml or 4 tsp of the anionic polymer, a dose of about 20 mg/L; the solids should collect into 1/8 to 1/4-in. particles
- Stop mixing and allow the solids to settle, they should settle more quickly than without polymer addition
- Pour the water out of the beaker or jar, leaving the solids behind
- The decanted water represents the treated water, and the solids are the treatment sludge

If you successfully followed the recipe you should end up with clear water and a small amount of sludge or slurry left behind in the treatment jar. You could have decanted the treated water without adding polymer to the solution. Try it that way. You should find that the solids are lighter, easier to disturb, and not as easy to leave behind.

After you have learned how to do it with ferric chloride and lime, try it with other treatment chemical combinations, if you are so inclined. Ferric chloride and caustic, aluminum chloride and lime or caustic, and ferrous sulfate and lime or caustic are possible combinations. Start with the same doses as used for the ferric chloride and lime combination.

Run an Oily Water Blank

If you are treating oily water or a turbid solution, you can mix up emulsified oily water to check out the chemical precipitation method. To make the oily water, prepare 500 ml of a detergent solution per the manufacturer's instructions and add seven drops (about 0.2 g) of motor or other petroleum oil. The oil should disperse in the water, causing it to become cloudy. The solution now contains about 0.4 g/L (400 mg/L) or 1.5 g/gal of fats, oil, and grease (FOG). This is well over allowable discharge limitations.

Now treat the prepared oily water using the recipe given above. You should again end up with clear water and a well-defined floc. The results of treatment of an oily water blank are shown in Figure 10.

Try It on Your Wastewater

Chances are that your wastewater contains emulsified oil or finely dispersed solids, such as found in paint booth wash water, or both, such as found

Figure 10 Oily water bench test results. The appearance of the **U** and **T** behind the beakers indicates the clarity of the water. The **U** behind the untreated water is less visible than the **T** behind the treated water.

in caustic hot tank rinsewater. The water is probably turbid or cloudy. Try the following:

- Place 500 ml or 16 oz of the wastewater into a beaker or jar
- Record the pH
- Add 10 ml or 2 tsp of 5% ferric chloride solution
- Test the pH, if it is not below 5, then add 5% acid solution until it is and record how much acid it takes; in the future, add the acid *before* the ferric chloride
- The amount of acid needed will depend on how much base or alkalinity is in the wastewater (rinsewater from a caustic hot tank may contain 0.5% caustic; in this case about 50 ml or 2 oz of 5% acid solution would be needed to bring the pH down)
- Mix the wastewater for about 5 min; you may see small particles at this stage
- Add lime until the pH is about 8.5; record how much it takes
- After you have adjusted the pH to about 8.5, stop mixing the solution; solids should slowly settle
- Gently mix the solution and add 20 ml or 4 tsp of the anionic polymer
- Stop mixing and allow the solids to settle
- Pour the water out of the beaker or jar, leaving the solids behind

Chemical treatment can produce relatively clear water. If you can produce clear water, you have an indication that the treatment worked. However, the clear water may still contain metals if you have not brought the pH up high enough, or if a compound is in the water that keeps the metal dissolved in the water. Remember, out of all the variables associated with chemical precipitation, the pH of the wastewater is the most important. Coagulant doses and mixing efficiency are factors, but are usually less critical than pH.

If the water is still turbid after treatment, the method did not work as well as it could have, and you should review the treatment recipe to see if you followed it accurately. If you followed the procedure without good results, then make sure the pH dropped to below 5 during treatment. If the pH does not fall dramatically when the ferric chloride is added then add acid first. Try adding twice as much ferric chloride, or try treating the same batch of wastewater again following the same steps. Sometimes double treatment is needed to produce acceptable effluent. You can experiment with doses and chemical addition sequences to determine what works best for your particular wastewater.

Try other treatment chemical combinations on the wastewater following the same procedures as you did with the ferric chloride and lime. Record what the treated solutions looked like. How clear are they, and what did the settled solids look like? Much is revealed just by looking. If the recipe works well, then you should end up with clear water and a small amount of sludge or slurry left behind in the treatment jar.

Select the chemical combinations that produce the clearest water and repeat the tests with them to see if the results are reproducible. Then take the treated water from the best-looking combinations and have your laboratory confirm that the wastewater will meet discharge limitations. Compare the treated water results with untreated wastewater to verify that the untreated water had significant amounts of contaminants in it at the beginning.

Collect at least two other samples of wastewater and bench test the best chemical combinations to verify that the recipe will work. Take care to ensure that the samples you collect represent the dirtier side of your wastewater and contain the contaminants of concern. Do not conduct your testing on the cleanest wastewater you produce, but do not expect the treatment method to be able to deal with everything. If your wastewater varies greatly during a shift, the best approach is to collect all of the water from a shift and test a sample of it. Alternatively, you can grab samples during the day, mix them together, and test the composite. Review Chapter 2 for sampling methodology.

The treatment method *may* work on the absolutely worst wastewater you can produce. If it does, you have a great method. If it does not it does not mean that the method will not work. Typically, you will collect wastewater for treatment and mix the worst wastewater with cleaner wastewater. Therefore, bench test a blend of wastewater. This is allowable by law if the wastewater is generated by the same process or if the wastewater must be diluted for treatment. It is not legal to dilute wastewater merely to make it meet discharge limits.

Bench testing can be used to screen alternative treatment methods. It is a relatively inexpensive way to gain information about the applicability of a particular technology. You can do a range of tests much faster than if you use a large piece of treatment equipment. Bench testing and what you learned from it can be used to troubleshoot your system after you have installed it.

Do a Pilot Test

Conduct a pilot test after your initial screening of treatment technologies. You have read, visited other shops, talked to suppliers, and conducted the necessary bench testing. The next step is bringing in equipment to see if it really works in your shop.

The purpose of a pilot test is to evaluate the treatment method under field conditions before investing in a permanent full-scale facility. If the test is conducted under realistic conditions, it can provide good insights into the pros and cons of a given technology, and it can provide needed design information. Whenever possible, you should conduct a pilot test before buying your system.

A fraction of the wastewater stream is treated in a typical pilot test. The equipment is smaller than required to treat the entire wastewater stream. The pilot test equipment should duplicate as closely as possible the function of a full-scale unit, but it can require more operator involvement and attention because it is designed to demonstrate the technology at the lowest possible cost. A supplier does not want to tie up capital in test equipment.

Suppliers that provide pilot testing should have enough experience with the equipment to be able to predict how well a full-scale system will perform based on pilot results. In some cases the pilot tests can be applied directly to the final design. For example, chemical types and dosages needed for chemical precipitation are usually the same for a small or large batch. The required ultrafilter membrane area is proportional to the wastewater volume. A test with short tubular membranes is predictive of the efficiency of longer units. In other cases the results must be extrapolated. For example, a larger clarifier, designed for 100 gal/min, may be more effective than the pilot unit, designed for 1 to 4 gal/min, and therefore clarifier size is not proportional to flow.

Small businesses usually do not produce much wastewater. Many pilot systems available from equipment suppliers are actually large enough to treat all of the wastewater from a small shop. In these cases the pilot system is a full-scale unit. You may end up doing a pilot test with a unit that is large enough to meet your needs. However, suppliers of equipment designed for larger flows (50 gal/min and up) may not even want to work with a small business.

If you obtain test equipment from a vendor, run it according to their instructions. The vendor may be willing to show you how to operate it. Make sure that you run the equipment yourself. Keep it in your shop and run it without assistance from the vendor for a few weeks. Treat enough batches of wastewater to become familiar and comfortable with the equipment. See if you can cope with problems as they develop.

If you decide to develop a system on your own, you must figure out how to pilot test it yourself. The following is an example of a chemical precipitation pilot test. If you develop your own treatment system for emulsified oily water or finely divided metal particulate-contaminated wastewater, you are likely to use chemical precipitation.

Figure 11 A 55-gal drum pilot test setup.

Pilot Test Example

Batch chemical precipitation can be tested readily using very little equipment. A 55-gal drum is an acceptable treatment tank and a 2 × 4 in. board makes a good mixer. You could also use a 5-gal bucket and a paint mixing stick, but a 55-gal drum will help you to get a feel for what it is like to treat larger batches of wastewater. Figure 11 shows a typical setup.

Metro South Facilities, Seattle, WA, a transit bus maintenance facility, generates oily water from the cleanout of oil water separators at commuter parking lots and from maintenance activities. Vacuum trucks are used to educt the oily water from the oil water separators. The water was being taken to an off-site treatment company at a cost of about $1.25/gal or $2,500/truckload. The decision was made to investigate onsite treatment to reduce waste management costs.

Through testing, the trucks were found to contain about one third settled solids. The truck operators, working with the pollution control manager, constructed a device to pump the water off the top of the solids in the trucks and into an existing holding tank. This action reduced the hauled waste and its disposal cost by about two thirds. The waste treatment company's charge was based on an assumption that the truckloads were primarily solids, without discriminating the water, which was less expensive to dispose of. If the waste treatment company had charged on the basis of solids content, the disposal costs may not have dropped as much. Oily water from bus maintenance activities was placed in the same holding tank because it had similar characteristics.

The oily water was bench tested and the following procedure was found to produce water acceptable to the sanitary sewer:

- Obtain the wastewater samples from beneath the floating oil
- Start with 500 ml (16 oz) of wastewater
- Add 0.5 g of aluminum sulfate (10 ml 5% solution)
- Mix the wastewater for several minutes
- Add lime to pH 8.5 (about 10 ml 5% slurry)
- Mix for several minutes
- Add 10 ml of 0.1% anionic polymer
- Mix gently until large floc forms and let settle
- Decant the treated water

The doses of treatment chemicals needed for 50 gallons of wastewater were then determined. To figure the chemical doses needed for the pilot test, the pilot test volume in gallons was divided by the bench test volume in gallons to obtain the volume ratio factor. In this example, the bench test used 16 oz and the pilot test 50 gal of wastewater; 16 oz, or about 500 ml, is one eighth (0.125) of a gallon; 400 times more chemicals are needed for 50 gal than for 0.125 gallon:

$$50 \text{ gal}/(0.125 \text{ gal}) = 400 \text{ or } (50 \text{ gal} \times 3850 \text{ ml/gal})/500 \text{ ml} = 385 \qquad (2)$$

The exact chemical doses are not as important as being consistent. Variations of 10 to 20% in aluminum or polymer dose should not adversely impact the process. For example, if you add 1 gal of 5% aluminum sulfate solution, instead of 1.3 gal, you are short 2 oz of aluminum sulfate. This should not make much difference in the effectiveness of the treatment. If a small variation in dose causes the treatment process to fail, then you should consider trying another method. In practice, doses will vary because the exact amounts of water and chemicals added probably will change slightly from batch to batch. One of your goals should be to put in a system that can tolerate the expected variations in chemical doses and wastewater contaminants. Do not act carelessly, but do not worry about getting everything exactly correct. You should worry about doing things consistently so that when something does not work as expected, you have a chance of figuring out why.

If you use the same bench and pilot test volumes in all of your tests, you will only have to go through this exercise one time. Then you can use the established factor to adjust your volumes. I prefer to use 500 ml or 16 oz in bench testing and 50 gal in pilot tests with the volume factor of 400. If you are consistent, it will be easier to keep track of things.

Next, multiply bench test chemical doses by this factor to determine the amount of chemicals to add to the pilot batch. The aluminum dose was 0.5 g/500 ml wastewater. For 50 gal, 200 g, or about 8 oz, is needed

$$0.5 \text{ g}^* \ 400 = 200 \text{ g} \tag{3}$$

and because 1 lb is 455 g, this is a dose of about 0.5 lb/50 gal. In the pilot test the aluminum sulfate could be added dry, but it is better to mix it with water first to ensure that it is mixed well into the wastewater. A 5% aluminum sulfate solution contains 50 g/L or about 6 oz/gal of aluminum sulfate; therefore, 4 L (or about 1.3 gal) of solution are needed for 50 gal of wastewater. The lime and polymer doses were calculated accordingly. For the pilot test, the following recipe resulted:

- Pump 50 gal of the wastewater from beneath the floating oil in the holding tank into a clean open-top drum
- Take a "before" sample
- Add 200 g of aluminum sulfate (4 L or 1.3 gal 5% solution)
- Mix the wastewater
- Check the pH to verify that it is below 5
- Add lime to pH 8.5 (about 200 g or 4 L 5% slurry)
- Mix
- Add 4 g anionic polymer (4 L of 0.1% solution)
- Mix gently until large floc forms and let settle
- Decant the treated water
- Obtain samples for laboratory confirmation of effectiveness
- Measure the sludge volume

Pilot tests were conducted using several different batches of the oily wastewater. Results are shown in Table 19. Several different batches of wastewater were tested. The treatment method was found to reliably produce wastewater that would meet the FOG discharge limitation of 100 mg/L.

It is difficult to precisely judge the volume of sludge that will be produced in a full-scale unit from pilot test results. The sludge can be filtered and weighed, or the volume can be measured, to estimate how much sludge may be produced in the full-scale unit. In this case the sludge volume was observed to be about 1 in. on the bottom of the drum, or about 1.5 gal. This equates to about 40 gal of sludge from a 1,400-gal batch. The full-scale unit actually

Table 19 Example Pilot Test Analytical Results

Parameter	Untreated conc. (mg/L)	Treated conc. (mg/L)
FOG	1200	35
	1600	20
	2100	40

produces about 70 gal of sludge per batch. The 70 gal of sludge includes some water because the water cannot be decanted completely from the treatment tank without pulling sludge into the effluent.

Sludge resulting from chemical precipitation is typically about 4% solids or 96% water. Because the sludge is mostly water, its volume can be reduced greatly by dewatering it through filtration or drying, lowering disposal costs. The pollution control manager at this facility decided to get the treatment system up and running before developing sludge dewatering procedures. Therefore, no sludge dewatering testing was done at this time.

When conducting the pilot test, record the steps you take and the potential difficulties involved in scaling up the treatment process. Remember, if you are pilot testing 50 gal and anticipate installing a 1000-gal system, you will be dealing with 20 times the treatment chemicals and 20 times the sludge. You will be mixing a larger volume of water, and thus a larger mess to clean up if the treatment does not work.

In this pilot test the treatment method produced wastewater of acceptable quality for discharge to the sanitary sewer. The treatment procedure could be explained to and understood by the operators. The equipment needed for a full-size system could be assembled at the facility. The resultant sludge could be put into the vacuum trucks to be disposed of with the untreated solids left in the trucks.

A pilot test may be the single most important thing you do when designing a pretreatment system. You may read that a method works, or a method may be working just fine in another shop similar to yours, but your wastewater may be different enough from the other wastewater that it must be treated by a different method. Until you actually try something for yourself on your own wastewater you cannot be sure that it will work.

6 SELECTING THE BEST ALTERNATIVE

INTRODUCTION

This chapter considers the selection of an appropriate wastewater management alternative based on the particular needs of your shop. It reviews selection criteria and explains how to do a cost benefit analysis. When searching for the alternative that, although it may not be the least expensive, has the highest probability of keeping you in compliance with the regulations, use the following criteria in the selection process:

Is it effective? Does it produce acceptable wastewater?
Can you afford it? What does it cost to purchase and install? What does it cost to run and maintain?
How much training and expertise is required to run and maintain it? Can you and your operators handle it?
What kind of waste treatment residual is produced? How will it be managed? What will it cost to dispose?

Each selection criterion involves many factors. Therefore, you need to be systematic in your selection process. You do not want to evaluate every possible alternative because of the vast number of them, and you will not have time to thoroughly analyze all of them. What you need to do is a gross assessment of technologies to eliminate those that are obviously not applicable to your wastewater, cost too much, or require too much attention or training to run. The alternatives selection process is outlined in Table 20.

FIRST SCREENING — ROUGH CUT

Use a step-by-step procedure. Start by characterizing your wastewater, then obtain information on wastewater treatment systems. Determine if a particular technology is used to treat wastewater similar to yours. Use the references in Chapter 9 to help. Talk to other shops, vendors, and regulators to get an idea of

Table 20 Alternative Selection Outline

Include in your selection criteria
- Does it produce acceptable treated water?
- Is it within your budget?
- Can your shop operate and maintain it?
- What waste treatment residuals are produced and how are they managed?

Compare the cost and benefits of each alternative
Select the most appropriate alternative

what is used. Ask for wastewater characterization data, which includes not only a chemical analysis, but a description of the processes that generate the wastewater. Ask for treated water data, and make sure that the tests they did are equivalent to the tests required by your pretreatment authority. Ask how long a given system has been in place and what problems have been encountered. Also, find out why they think the system is working as it should. Has the pretreatment authority taken a look at the facility and found it satisfactory?

Make a list of available alternatives that may be feasible for your shop. Specifically, include treatment systems that are used by shops similar to yours. For example, another machine shop may be using an evaporator to concentrate 50 gal/day of hot tank rinsewater, or a radiator shop may be using chemical precipitation to treat their wastewater.

Eliminate from consideration those alternatives that are not applicable. For example, ultrafiltration cannot remove soluble metals, an oil water separator will not effectively treat emulsified oily water, and ion exchange is not suitable for oily water. Do some homework now to narrow the options you consider. What you learn will also help you to evaluate the options that look viable, because you will understand more clearly why they are applicable. Document your reasons and move on with your shorter list.

Next, look at the approximate costs of treatment systems. Gather quotes from vendors. Estimate the cost of installing a chemical precipitation system yourself. The information can help you decide how to proceed. A variety of ways exist to use the various treatment technologies and the approach chosen affects costs. Typically, the more automated the system, the more it costs. Different vendors may offer similar technology at very different costs because they offer different services. At this point rule out systems that are obviously over your budget. After you learn more you may reconsider some of the options that you rule out, but for now the object is to narrow down the options so that you can do a more detailed evaluation of a few.

SECOND SCREENING — A CLOSER LOOK

Does It Work?

At this stage you have determined that certain technologies may be applicable to your wastewater, but will they be effective in your shop? Obtain treatability test data, using your wastewater, for each alternative. It is very important not to rely on vendor claims — you must gather your own data. A vendor can help you in this regard, but ensure you are involved in the process.

Conduct bench tests in your facility or have a vendor do a treatability test. Pilot test the systems that perform best in bench testing, or that seem the most cost effective. You may be able to rule out certain alternatives based on the treatability bench test data.

Consider the flexibility of the system. If your wastewater characteristics change because of process or work load changes, can the process be adapted to treat the wastewater? Can it be run infrequently if your wastewater flow is reduced by water conservation or because of a smaller work load? Can it treat more concentrated wastewater if reducing rinsewater flow increases contaminant concentrations?

What Does It Cost?

The next step is to do a cost analysis for the systems that performed well in the bench and pilot tests. Many costs are associated with a wastewater treatment system (see Table 21). You must identify the most significant costs of each system under consideration. Develop preliminary designs for each wastewater management alternative that you evaluated and found to be potentially feasible, and, at a minimum, estimate the following costs for each of them:

Purchase and installation
Operation and maintenance
Disposal of treatment residue

Bench and pilot testing cost are part of the upfront project costs. They are counted toward the total project cost, but should be excluded from the final cost analysis.

Permit application preparation and permit fees should be determined so that you will know what to expect. The wastewater discharge permitting process will be similar for each of the alternatives. An air emissions permit, if needed, will add to the permitting costs. Also determine the costs of building, plumbing, electrical, and fire department permits, if they are required. Reports may be required by each of the permitting authorities. Determine whether they will accept a report that you prepare or if you must hire an engineer to prepare the report. Estimate the cost of report preparation. The permitting costs may be similar for each alternative. If they are not, then use this information to help select the alternative.

Purchase and Installation

What will it cost to purchase and install the system? Based on the preliminary system design, which should show the major system components, prepare a parts list. Estimate the cost of the major system components and the labor

Table 21 Budget Items Associated with the Implementation of a Wastewater Treatment System

- Design
  - Testing
  - Selection of equipment
  - Specification preparation
    - Sizing of equipment
    - Operating parameters
    - Drawing preparation
- Permitting
  - Report preparation
  - Fees
  - Meetings
- Equipment procurement
- Installation
  - Assembly of equipment
  - Shop modifications needed
    - Utilities
    - Berms
    - Ventilation
    - Storage
- Documentation
  - Operation and maintenance manual
- Training
- Operating and maintenance
  - Labor
  - Equipment upkeep
  - Utilities (electricity, water, air, gas, steam)
  - Disposables (chemicals, filters)
  - Wastewater analysis
  - Regulatory records and reporting
  - Discharge fees
- Residual disposal

needed to build and install the system. You will need the estimates to select the best alternative for your shop.

Of course, the cost and labor associated with building the system depend on how much you do yourself and where you purchase the equipment. You can get quotes from vendors for packaged plants and for system components. If you buy a package system, then the cost of system construction is included in the vendor quote. If you put together a system yourself, it is likely that you will obtain the components and assemble them and hire electricians, plumbers, welders, and carpenters when they are needed. You may have tanks, mixers, and pumps that you can refurbish and use, or you may buy used equipment. Whether you do the work or hire someone to do it, you need to get a rough estimate of the cost of the assembled system.

Estimate the costs of facility modifications that are needed to install each alternative. You will need to provide a concrete or asphalt pad and berm for any treatment system. The cost will depend in part on the size of the system footprint. You will need to collect and transport the wastewater to and from the system. The utilities required will depend on the system. Determine what the electrical, gas, water, compressed air, and ventilation requirements are and if you already have the necessary hookups, or if you will need to obtain them. You may have room in your shop for the equipment or you may build an addition. If you must put the equipment outside, then, at a minimum, build a roof over it to keep out rainwater. It is best to enclose the treatment system to protect it and the operators from the elements. Think through what other major modifications may be required.

Operation and Maintenance

The treatment equipment and the labor needed to run it are typically the highest cost items. Disposal of the treatment residue also can be significant. Treatment systems can require constant attention to run or can operate virtually unattended; there is a wide range of operational requirements. When evaluating alternatives you need to take this into account. You choose between a system that will cost more to install, but will require less attention to run, and an inexpensive system that will take more time to operate.

The decision depends in part on how much wastewater you are treating. If only 50 gal/day are treated then you may be reluctant to spend $10,000 for an automated system if you can spend $500 for a manual one that takes 1 hour/day to operate.

What does each alternative cost to run and maintain? What will the utilities cost? Utility costs can vary significantly, depending on the operating principle of the treatment system. Utility costs are not usually a factor in small treatment systems. Electricity costs will be negligible for a chemical precipitation unit with small pumps and mixers. For example, electricity for a 50 gal/day evaporator may cost $5/day, a small fraction of the operating and maintenance costs.

How many hours per day does it take to run the system? You will need to dedicate someone's time to the system. Do not treat it as an afterthought. Give it priority and arrange the workload and budget to accommodate it.

Consider the cost associated with training. What initial and ongoing training is required? Are training classes available, and if so, what is the cost? What kind of background does the treatment system operator need? Will you need to hire someone with the appropriate experience to run it, or can you or one of your current employees handle it? The level of expertise needed will vary with the type of treatment system.

Estimate routine maintenance costs for parts and labor. What routine maintenance does the equipment need? Do some items need periodic replacement? For example, ultrafiltration membranes may need to be replaced every year or two at a significant cost. Cartridge filters need frequent replacement. Pumps and motors should get routine preventive maintenance. Ask the equipment supplier or someone who has used similar equipment for 1 year about maintenance requirements.

Disposal of Treatment Residue

What kind of waste treatment residual is produced by each alternative and how will you manage it? Using pilot test results and information from similar shops, estimate how much residue you will produce and characterize it. Find out if the residue has enough value to be recycled. Some sludges contain enough metals to make it cost effective to recover them.

Contact the appropriate disposal facility and determine their requirements for accepting the waste and what they will charge. They may require periodic analysis of the residue, which you should also include in your budget. Determine the storage and handling requirements. Do you need to build a storage area? How much time will you spend handling the residue? The major costs associated with residual management are analysis and disposal.

Operating Requirements

Different technologies and applications require different levels of expertise to operate. You must decide whether learning how to run and maintain a system that is inexpensive but unfamiliar is worthwhile. For example, you may be able to put in a chemical precipitation unit for 100 gal/day for $1,000, while an evaporator may cost $3,000 and an ultrafiltration unit may cost $5,000. You can learn about chemical precipitation and reduce your capital costs, but you may not like it enough to put forth the effort. The benefit of going with evaporation may outweigh the initial cost.

Another factor to consider is the nature of the chemicals needed for treatment. For example, chemical precipitation probably will require the use of acid and base. If you have a caustic tank you will need plenty of acid to neutralize its rinsewater. You must deal with safety and health issues associated with the chemicals. Find out what you are getting into and determine if you are willing to live with the consequences.

Treated Water Disposal

Examine your disposal options. Determine which treatment system alternatives produce wastewater that is compatible with each method of disposal. For example, consider what you are willing to pay to keep your wastewater out of the sewer. You may not have the option of discharging to a sanitary sewer or may not want to deal with the potential liability of an out-of-compliance discharge. In this case you will be willing to pay more for a system that can get you out-of-the-sewer.

Waste Treatment Residual Disposal

The nature and volume of the residual depends on the type of technology you use. Ultrafiltration and evaporation produce wastewater higher in contaminant concentrations but lower in volume than the original waste stream. Chemical precipitation produces a slurry that you can filter to reduce in volume. Ion exchange regeneration produces concentrated metal-bearing wastewater. Activated carbon concentrates organics.

The residual may be oily water that can be disposed of as waste oil. The residual may be hazardous waste. Consider the hazards and costs of residual handling and disposal.

Do It Yourself or Buy a Packaged Plant

A range of options for development and installation of a treatment system is available. You can do everything yourself or have a vendor or consultant do everything, or split the tasks between yourself and a vendor or consultant.

The advantage of doing it yourself is that you will necessarily learn about each aspect of treatment system implementation and, therefore, will be self-sufficient once your system is up and running. You can more readily adapt the system to changing needs and requirements. You can potentially install the system at a lower capital cost because you are not paying for someone else's overhead, and you can customize the system to your operation. On the other hand, you must spend more time learning about the technology. You will need to read and experiment. If you are not inclined to do this, you obviously must pay someone else for their expertise.

You can buy a packaged plant and have the vendor install, start up, and train you to run it. This method has the advantage of being the least time consuming on your part. If the vendor knows what he/she is doing, you benefit from this experience and the system has a high probability of working. The disadvantage is that it may be more expensive than the system you would build for yourself, investing time instead of money, and you risk discovering that the technology is not appropriate to your situation and will not work because the vendor was a salesperson without much technical expertise.

Benefits and Deficiencies

List the benefits and potential deficiencies or problems associated with each alternative. You may need to change your processes or workflow to make the treatment system work. How will this impact the productivity of your shop? Will the processes changes have a beneficial impact by reducing employee exposure to potential hazardous substances and by reducing waste generation?

How long will it take to put the system in? Do you have a regulatory deadline? It will take time to design, install, and start up a system. An off-the-shelf system may be available for quick installation, or a vendor may need 3 months lead time to get a system to you.

If you are buying a system from a vendor, what level of service does the vendor supply, and are other customers satisfied with the vendor's support?

Can you understand the principles of operation of the system? Can the system be started up and shut down easily? Can you fix it yourself? Are parts and chemicals readily available or are you dependent on a single source?

What operational hazards are associated with the equipment? For example, are acid or caustic used, dust produced, ozone generated, or ultraviolet light present? Are you willing to live with those hazards? Are you willing to handle the treatment chemicals and the waste treatment residue?

Is the system designed to minimize the chance of inadvertently discharging untreated wastewater? A continuous system will need more safeguards to prevent out-of-limits discharges than will a batch system. With a batch system you can test a batch before you manually open a valve to release it to the sewer. A continuous system could be designed to discharge into a holding tank for testing prior to discharge.

COST BENEFIT ANALYSIS

A cost benefit analysis is an objective assessment of the advantages and disadvantages of alternatives in relation to the corresponding costs. To do one, list the remaining alternatives and their capital, operating, and maintenance costs. Figure the cost savings associated with each alternative as compared to a baseline cost such as sending the wastewater off-site for treatment. Cost analyses are included in the wastewater treatment system examples presented in the Appendices.

The treatment system must produce treated water that meets discharge requirements. Eliminate the alternatives that do not. A monetary fine is one tangible drawback of not meeting discharge limitations. Trouble with local regulators and bad public relations are less tangible potential drawbacks. The benefits of installing a treatment system outweigh the cost savings and difficulties of not doing so, or you would not have read this far.

You need a treatment system, but your budget is limited. Therefore, eliminate the alternatives that are outside your budget. Any potential benefits associated with those alternatives are outweighed by the fact that you cannot afford them. You will need to "make do" with a less expensive option.

You are left with several alternatives which work and which you can afford. Now you must consider how your shop operates, what type of processes and equipment you already have, and how much time you must devote to the development and care of a treatment system. These factors will help you do your cost benefit analysis.

MAKE YOUR DECISION

Examine your situation. Consider the advantages and disadvantages of the various alternatives you have evaluated. Include intangible benefits and risks. Determine how much you are willing to pay for the things that you want and select the most appropriate system for your shop.

7 IMPLEMENTING THE WASTEWATER TREATMENT SYSTEM

INTRODUCTION

After you have selected the alternative that best fits your shop's needs you must install it and start it up. Then you will be faced with the tasks of operation and maintenance. To implement the system, the following must be accomplished:

- Document the design basis
- Prepare the final design
- Prepare an operations and maintenance manual
- Obtain the necessary permits
- Train shop personnel
- Install the equipment
- Start up and evaluate the system
- Operate and maintain the treatment system

The implementation steps are presented in Table 22. How elaborate these tasks will be depends on how complex the treatment system is and how much scrutiny it will get from the regulators.

Table 22 Wastewater Treatment System Implementation Steps

Characterize your wastewater
Evaluate wastewater management options
Design the facility
Prepare an operation and maintenance manual
Obtain the necessary permits
Train shop personnel
Install the equipment
Start up and test the system

MINIMUM DOCUMENTATION REQUIREMENTS

Your ability to demonstrate that you are in compliance with the applicable regulations is your responsibility. Therefore, you should document your waste management procedures, including the purpose and effectiveness of your treatment system. At a minimum document the following:

A description of your shop processes
Wastewater characterization
Why the system was installed
How you know the system will produce acceptable effluent
A drawing of your shop, showing the location of the processes in relation to the treatment system
A drawing of the treatment system, showing the process tanks, pipes, pumps, mixers, and other equipment, and the flow of the wastewater through the system
The drawings should also include the location of floor drains, sewer connections, and spill prevention devices, such as berms
Operation and maintenance procedures

What you will need to convince a regulator that your system is effective and to build the facility depends on how complicated the system is. If you have a small wastewater discharge volume which does not require a permit, you will need less documentation than for a large permitted discharge. If you are building a $100,000 system to treat 50 gal/min, you will generate much more paperwork than for a 100-gal batch rinsewater treatment tank.

The design basis report, plans and specifications, and operation and maintenance manual are described in detail in the following sections. When reading those sections, keep in mind that the descriptions apply to a permitted wastewater treatment facility. If you are installing a small facility that does not require a permit, fulfill the minimum requirements and include whatever else is asked for by your pretreatment authority. The wastewater treatment system examples, presented in the Appendices, describe the documentation needed for implementation. The examples can help you decide how to proceed.

DOCUMENT THE DESIGN BASIS

Prepare a report that documents the system design. The purpose of this is to document the work done to develop or redesign the waste treatment system. The report should be complete and detailed enough so that a person familiar with waste treatment can assess whether an adequate job was done in selecting the waste treatment process chemistry and equipment. The report should contain, at minimum, a concise summary of the characteristics of the waste stream, a description of the procedure used to select the treatment process, and a description of the treatment process. The report can be used to organize your thoughts, train operators, and respond to questions from regulators. If you can

Table 23 Guidelines for the Preparation of the Design Basis Report

The purpose of a treatment system design basis report is to document the work done to develop or redesign a pretreatment system. The report should be complete and detailed enough so that a person familiar with waste treatment can assess whether an adequate job was done in selecting the wastewater treatment process chemistry and equipment. Check with your pretreatment authority to determine their specific requirements. It should contain at minimum the following information:

- Type of industry or business
- Kind and quantity of finished products
- Volume and characteristics of the wastewater and how it is generated and disposed of, including: spent process solutions, process rinsewater, noncontact cooling water, and evaporation, if significant
- A site map showing the location of the pretreatment system
- A layout diagram of the pretreatment system to include the location of wastewater sources at the site, the routing of wastewater, and the sewer discharge point
- A description of the physical provisions for containment of oil and hazardous material spills and prevention of accidental discharge of untreated wastewater
- Sound justification through the use of bench testing, pilot plant data, results from other similar installations, and/or scientific evidence from the literature, that indicates that the treated water from the proposed facility will meet applicable wastewater discharge permit limitations
- Basic design data and sizing calculations of treatment system components (e.g., pumps, tanks, mixers)
- A description of the treatment process, including the amount and kind of chemicals used in the process
- A flow diagram of the treatment process
- A discussion of the treatment residual disposal method
- A statement regarding compliance with your state's environmental policy act, if applicable (this could require an environmental impact statement)
- A schedule for final design and construction

provide the information outlined below, you have done a thorough job and have increased the probability that your treatment system will work.

Guidelines for the preparation of the design basis report are given in Table 23. Describe your facility and the wastewater it generates. Provide wastewater characterization data. Include diagrams of the site showing the location of the treatment system, wastewater sources, wastewater routing, and the sewer discharge point. An example site diagram is shown in Figure 12. Show where the wastewater ends up and outline the spill containment structures. Describe the physical structures for spill control and accidental wastewater discharge prevention. The drawing(s) does not have to be elaborate for a simple facility. It must clearly show the shop layout. It is a map of your shop to help others understand how it is set up, what processes you have, and how the sewer discharge point is protected.

Present the results of your evaluation of treatment alternatives. Indicate the alternatives you considered, why you selected the one you did, and why you ruled out others. Document the process that you went through. Outline the steps you took and the decisions you made. Include your observations of treatment systems at other shops and any data that you received from them. Provide the data that you collected that show that the chosen treatment system will produce

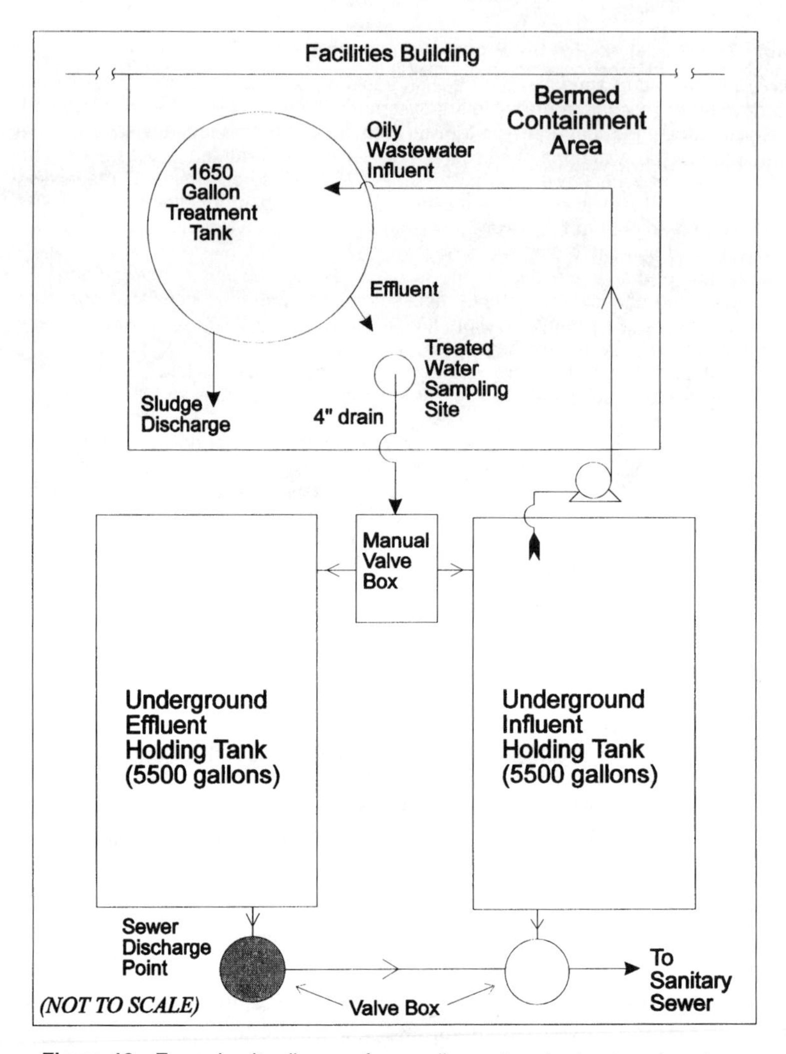

Figure 12 Example site diagram for an oily wastewater treatment system.

wastewater that meets discharge limitations including bench and pilot results. It is important for you to describe clearly the results of your evaluation. Lead the reader through the evaluation, providing enough information to show you understand why the selected process will work in your shop.

Provide basic design information, including a description of the treatment process. Include a process flow diagram showing the treatment tanks, the connections between them, the chemical addition points, the process monitoring points, and the treated water and waste residual discharge points. Tell how the process works and what chemicals are used. Give your treatment recipe. Outline the methods and tests you plan to use to ensure that the process is working properly, and the safeguards that you built into the system.

Explain how you will manage the waste treatment residual. Provide any characterization data that you have. Tell how you will collect and store them. Outline the residual disposal methods that you considered. Explain why you selected the particular disposal method.

Ask your pretreatment authority if an environmental impact statement is needed. Some pretreatment authorities have prepared environmental impact statements regarding wastewater discharges to their systems. They do this in conjunction with local limits development. If your pretreatment authority has prepared such a statement, the discharge of your wastewater may be covered. However, you may still need to prepare an environmental impact statement for the rest of your facility.

Provide a schedule for final design and construction. Show the major tasks that will be accomplished. Include the dates you expect to finish the final design and plans and specifications, install the system, receive approval to discharge, complete staff training, and start up the system. You may have some flexibility in the schedule if you are installing the system voluntarily. The schedule may be timed to meet a regulatory compliance order. If so, make sure you notify the regulators of any delays as soon as you are aware of them.

PREPARE THE FINAL DESIGN

At this stage, you have chosen a particular system. You prepared preliminary designs to allow you to do a cost benefit analysis and select the best alternative. You have documented the basis of your decision. Now you must prepare a detailed design so that the system can be built and installed.

The design process includes sizing the equipment, using waste volume, information from similar shops, and bench and pilot test information; providing spill containment structures; complying with building and fire codes; and providing the necessary utilities. A critical aspect of the design is the treatment plant layout.

A plant layout shows the major system components drawn to scale in their expected locations. The layout helps the designer to visualize how much room is needed for the system, the spatial relationships of the components, and how much space is between them. It helps to ensure that the system can be efficiently operated and maintained. Enough space must be provided for an operator to have access to the equipment. Control panels, mixing tanks, and sample points must be readily accessible. Physical hazards must be minimized and controlled. Spend some time thinking about and preparing your plant layout. Do not install a system without one. It is much easier to move a berm, tank, or pipe on paper than after it is in place.

Prepare plans and specifications which explicitly call out the design requirements. The plans and specifications are used to ensure that the finished plant will be built to minimum standards, to obtain supplier quotations, to allow the construction manager to determine if the plant is built correctly, and to satisfy a regulatory agency that the plant will be built as intended.

The plans and specifications should be complete enough that a supplier can install the waste treatment system as intended by the designer. The level of detail supplied in the plans and specifications will depend on the specific project. If you are installing a simple system, such as an evaporator, a 55-gal batch treatment system, or a small ultrafiltration unit, the plans and specifications provided by the supplier may be sufficient. If you are putting together a 1000-gal batch or a 100-gal/min continuous treatment system, complete with chemical mix tanks and pumps, then you will need more than a shop layout and process flow diagram to install the system.

The amount of detail necessary also depends on who is installing the system. You may buy a packaged plant, in which case you inform the supplier of your discharge limitations and shop constraints, and the supplier provides equipment plans and specifications. You are responsible for ensuring the required utility connections and floor mounts, based on a plant layout and process flow and control diagram. If you install it yourself, you must prepare an equipment list and at least a rough layout, so that you can put the pieces together. If a contractor installs the equipment, a more detailed work plan is required to minimize confusion.

Guidelines for the preparation of the plans and specifications are given in Table 24. The plans and specifications should include a summary of the basic design criteria presented in the design basis report. Include a description of the treatment process, a process flow diagram, the process monitoring points, and the discharge points. Tell how the process works and what chemicals are used.

Table 24 Guidelines for the Preparation of the Plans and Specifications

The plans and specifications should be detailed enough that the system can be installed as intended by the person who designed it. The plans and specifications should include at minimum the following:

- A summary of the basic design criteria presented in the treatment system design basis report
- An explanation of any deviations from the original design basis in as much detail as was provided in the original report
- A description of the general pretreatment system operating procedures including startup, shutdown, and process control
- A general site drawing showing the location of the pretreatment system and related piping and utilities in relation to the plant site and process facilities
- A layout of the pretreatment system showing major components, including:
    - Spill containment structures
    - Influent collection system
    - Collection sumps and pumps
    - Treatment system components (e.g., tanks, pipes, pumps, mixers)
    - Sludge handling equipment
    - Influent and effluent piping
    - Foundations for major structures
    - Location of the treated water discharge point
- A schematic which includes process instrumentation and control devices (e.g., valves, pH controllers, flow meters, level controllers) in relation to the treatment plant equipment and piping, and sampling locations (in some cases, where the treatment process is relatively simple, these can be shown on the treatment system layout)

Outline the safeguards to be built into the system. You can refer the reader to the design basis report for details. Explain any changes that were made in the design.

Include a general description of the pretreatment system operating procedures. Detailed procedures will be provided in the operation and maintenance manual. Outline how the treatment will be started up, routinely operated, and shut down. Explain how the treatment process will be controlled.

Provide a general site drawing which shows the location of the pretreatment system and related piping and utilities in relation to the plant site and process facilities. This drawing is a large-scale map of your shop. Details of the pretreatment system are shown on the treatment system layout and schematic.

PREPARE AN OPERATIONS AND MAINTENANCE MANUAL

An operations and maintenance manual should be prepared before the treatment plant is started up to provide a guide for the plant operator. Revise the manual after the plant has been running for awhile and routine procedures are established. Plan to periodically review the manual and update it to include significant changes in plant operation.

The manual can be used to train your operators and provide a reference when problems occur. Also, it will be useful when the primary operator goes on vacation or moves on to another job and someone less familiar with the operation of the plant takes over. These transition periods are critical and you need to reduce the inherent risks by providing detailed operational instructions to the less-seasoned operator. The purpose of the manual is to present technical guidance and regulatory requirements to the operator, for both normal and emergency conditions. Guidelines for the preparation of the manual are given in Table 25.

The manual must contain the names and phone numbers of the people responsible for the management and operation of the system, regulatory reports, and emergency response; provide home phone numbers of those with overall responsibility; include a description of the design criteria, treatment process, and each system component; and provide a process flow diagram, plant layout, and a schematic of the process control instrumentation. This information is already included in both the design basis report and the plans and specifications. Both reports could be included in this manual if they are not voluminous; if they are, include only the most pertinent sections.

Describe normal operating procedures, step by step. Provide enough detail to prompt a trained operator. Explain the sequence of events used to turn it on, run it, and shut it down. Explain how to set and measure process control parameters, such as flow rate and pH. Explain the effect of the various process control parameters on plant operation and effluent quality. Tell what happens if the flow rate is too high or process time is shortened. If applicable, explain how critical pH is, and the importance of keeping free oil out of an ultrafilter

Table 25 Guidelines for Preparation of Operations and Maintenance Manual

The purpose of the manual is to present technical guidance and regulatory requirements to the operator for both normal and emergency conditions. The manual should include the following information:

- Names and phone numbers of the responsible individuals
- A plant process description, flow pattern, normal operating procedures, and expected treatment efficiency
- Principal design criteria
- A description of each plant component, including its function and relationship to other plant components
- A process flow diagram, plant layout, and diagram of process control instrumentation
- Explanation of the effect of the various control parameters on the operation of the plant, i.e., flow rates, process times, pH, sludge settleability, etc., and their effect on effluent quality, including a description of recommended control parameter settings
- A discussion of how the facilities are to be operated during normal conditions and anticipated startups and shutdowns to maintain efficient treatment
- A section on effluent monitoring procedures, including sampling techniques, monitoring requirements, and sample analysis; sample analysis procedures refer to the analytical methods used and whether the laboratory is in-house or an outside contractor
- Recordkeeping requirements and procedures with samples of the forms that are used
- A maintenance schedule with procedures incorporating equipment manufacturer's recommendations and preventive maintenance and housekeeping requirements
- A safety and health section, including a description of the potential hazards present, safety equipment provided, and the mandatory precautions the operators must take
- Emergency plans and procedures, including who to notify, and emergency shutdown, spill response, and evacuation procedures

and solvent out of an evaporator. Describe what the treated water and treatment residual typically look like.

Include effluent monitoring and regulatory reporting requirements. Specify who is responsible for them. Give examples of the data to be recorded and provide sample forms. Describe where, how, and when to take samples. Detail how to preserve and store the samples. If the operator is responsible for the analysis, provide the analytical procedures. If the samples are sent to another laboratory, provide transport procedures.

Provide a preventive maintenance schedule and procedures. The schedule may be organized by system component, process step, or by daily, weekly, monthly, and yearly requirements. Incorporate equipment manufacturer's recommendations. You may include their operation and maintenance manuals. Also, incorporate housekeeping requirements, including equipment and floor cleanup, chemical storage, and management of housekeeping waste.

Include a safety and health plan. Describe the potential hazards present and mandatory safety precautions. List the available safety equipment and tell where to find it. Give the location of the treatment chemical Material Safety Data Sheets. Include an emergency response plan. Tell who to notify and how to reach them. Explain how to conduct an emergency treatment system shutdown. Provide a spill response plan, including an evacuation plan, and spill containment and cleanup residue disposal procedures.

OBTAIN THE NECESSARY PERMITS

You may need to obtain a wastewater discharge permit to legally discharge the treated water. At the very least you should obtain verbal approval to ensure you understand the rules. A wastewater permit application will typically require that you document your wastewater characteristics, treatment plant design basis, plant layout, operational requirements, and sampling and monitoring procedures. The pretreatment authority will want the same information that you need to design, implement, and run the plant. They want to ensure that the system will reliably produce treated water that meets their limitations. This is a shared concern. If you have done a good job in installing your plant, you have what they want.

The design basis report, plans and specifications, and operation and maintenance manual preparation guidelines presented in this book are based on Washington State regulations (WAC 173-303) used by staff engineers to review permit applications. They represent a sound approach to treatment system documentation. Check with your pretreatment authority to determine their specific requirements for the report. You should approach them early in your design process to establish communication. Plan to meet with them several times to get feedback. Establish a schedule for submittal of the application documentation to facilitate your planning.

TRAIN SHOP PERSONNEL

It is your responsibility to ensure that you and your operators understand the operation of the treatment system well enough to run it effectively. Determine what training is necessary and prepare a training plan. Some regulations, such as the Federal Occupational and Safety and Health Act, require that you have a safety training plan and maintain a training record. Set up training sessions to cover operating principles and regulatory requirements.

Commit to good personnel and give them the time they need to do their job well. The operators should be interested in wastewater treatment and be motivated to produce good quality effluent. Let them take advantage of training opportunities at local community colleges. Allow them to attend local waste management conferences and workshops. They will gain a better perspective and can help your shop keep up with the regulations and technology.

Involve your operators in the design process before you build the system. Their sense of the function and limitations of the system will be better than if they are recruited after you install it. They can offer their advice based on their experience to help design a system that they can run and maintain.

Give the operators enough hands-on experience with the system that they can confidently operate it. Emphasize the prevention of out-of-limits discharges. Stress health and safety requirements and precautions and emergency procedures. Ensure they all know where to find and how to operate the shutoff valves and switches, the safety gear, and the spill control equipment.

Periodically make time for your alternate operator or operators to run the treatment system so that they will remember how to do so. Then, if your primary operator unexpectedly misses work, you will still have somebody in the shop that knows how to run the system.

Make sure that the rest of your staff understands what the treatment system is for and what wastewater it can handle. They may need to learn to change their practices to prevent treatment plant upsets. Everyone must understand and follow health, safety, and waste management requirements. Include the treatment facility in your safety training. Tell your staff what the potential effects are of dumping an incompatible hazardous material into the treatment system.

Establish clear lines of communication between the production staff and the treatment system operators and make sure the staff knows who is responsible for the treatment plant and what information they routinely need. Tell them what to do in an emergency situation.

INSTALL THE EQUIPMENT

Installing the equipment entails making the necessary shop modifications and obtaining and assembling the components. Some of the modifications such as preparation of the pad and its protective coating must be done before the equipment is set in place. Others, such as wiring and plumbing, can be done after it is placed. Plan the order of construction so that you do not have to rip something out or struggle to work around it. The treatment system is not your only worry. Do not forget about utilities, berms, ventilation, and chemical, treatment residue, and equipment storage.

Purchase and assemble the pretreatment system components. If you are installing a package system, have everything prepared to drop it in place when it arrives, especially if the supplier is coming onsite to start it up.

START UP AND EVALUATE THE TREATMENT SYSTEM

After you have installed the system, first check the functioning of all of the components that can be tested without adding water to the system. For example, ensure that mixers run and pH probes work.

Before you put any wastewater in the system test it out with tap water, if feasible, to check for leaks and proper functioning of the wastewater pumps and mixers. If you fill it with wastewater and something needs to be fixed, you must clean it out before you can repair it. Also, test the function of any chemical mixing equipment and pumps. If you are undertaking chemical precipitation, you may be able to produce solids in the system, by treating the tap water, to test your solids handling equipment.

After you have determined that the treatment system will hold water and mix it, introduce wastewater into it. Obtain a sample of untreated wastewater. Analyze it to verify, for the record, that it is representative of your wastewater.

(It is not necessary to analyze it beforehand unless you need the analysis to determine chemical doses. You can obtain the results along with the treated water sample.) Run a bench test, if you are doing chemical precipitation, to verify treatability and determine how much acid or base to add.

Set the treatment process parameters as noted in the operation and maintenance manual. Observe the wastewater as it undergoes treatment and compare your observations to what you have seen at other shops and in bench and pilot testing. A chemical precipitation process provides visual cues that an operator can use to determine if it is operating well. Adjust the process parameters as needed to make it look correct. Obtain a treated water sample. Analyze it and compare the results to your discharge limitations.

If you produced acceptable wastewater, you are in business. Keep a close eye on the next several batches. Obtain and analyze untreated and treated wastewater from them to verify stable plant operation. Provide the pretreatment authority with the data confirming a successful startup, and begin routine operation and maintenance of the system.

TROUBLESHOOT PROBLEMS

If you cannot produce acceptable wastewater, or if the treatment rate is too low, then you must troubleshoot the process. Run through the same procedures that you did with bench testing and try to reproduce the results. You may find that different treatment chemicals or doses work. If so, make the adjustments and use them. If you cannot reproduce the results you obtained when you did your initial treatability testing, then you did not obtain representative samples to design your system, or your wastewater characteristics have changed. In either case you have a serious problem and will need to work to modify the process or wastewater characteristics.

If you can produce results in a jar, but not in the treatment plant, examine each treatment process step. The most common causes of chemical precipitation treatment plant failures are incorrect chemical doses, poor mixing, and solids pass-through.

pH is critical, so ensure that the pH controller is working. Determine whether the chemical feed pumps are delivering the required chemical doses. If polymer is used for flocculation, make sure the dose is not too high. Too much polymer can cause the floc to become smaller and fail to settle well. An overdose of inorganic coagulant is not as serious as an overdose of polymer. An overdose of caustic, causing a higher than optimal pH, is more likely to result in poor metals removal than an overdose of lime. Lime-precipitated floc is more stable than caustic precipitated floc, and the upper limit of pH is about 12.5 with lime, but the pH can go over 14 with caustic.

Poor mixing in any stage of the process can result in ineffective floc formation and precipitation. It is easy to mix a jar on the benchtop. It is not as easy to mix a tank full of wastewater. Ensure that the mixers are doing their job.

Can they lift solids off the tank bottom? Does the water short circuit through the tanks? The inlet and outlet of the tanks should be on opposite sides of the tanks and one should be at the top and the other at the bottom. Round tanks should have baffles and/or off-center mixers to increase turbulence and to prevent the entire contents of the tanks from turning instead of mixing well. Use a side-mounted mixer for a large, round tank to provide mixing turbulence directed across the tank rather than a top-mounted mixer directed downward.

Make sure solids are not passing through the clarifier or being decanted out of a batch tank. Metals are concentrated in the sludge produced in the treatment process — this is how they are removed from the wastewater. Therefore, it is very important to ensure effective solids removal. If floc is carrying through your system, treatment performance will be compromised.

The most common cause of an evaporator failure is fouling of the heat exchanger. The evaporator will then boil off less water than anticipated and may not be able to keep pace with the wastewater flow rate. You may need to clean the heat exchanger more often than expected or you may need to buy an additional evaporator. Another cause of inefficient evaporation is stratification of the solution, preventing good heat transfer and inhibiting the vaporization of water. Oil on the surface can also reduce efficiency. The best way to avoid these situations is to ensure that the evaporator you buy can handle your wastewater and to size it to allow for less-than-optimum efficiency.

The flow-through ultrafiltration membranes will typically slowly decline in use. The membranes must be cleaned periodically to restore full flow. The cleaning frequency is a critical operational parameter. Cleaning generates wastewater that must be disposed.

Ultrafiltration membranes can be quickly coated with free oil, reducing the flow rate of permeate through the membrane. Ensure that free oil is removed from the wastewater before it enters the ultrafilter. The membranes can become irreversibly fouled and unresponsive to cleaning. For example, an ultrafiltration unit in a radiator shop was tested on test tank and oily flush water. Within a few weeks, the silicates in the antifreeze apparently plugged the membrane pores and could not be removed effectively by backflushing. However, a short-term ultrafiltration test may not detect membrane fouling problems that take time to develop.

Also, ask your equipment supplier for help. Talk with other treatment system operators to find out how they overcame the problem. The limitations of a treatment process are sometimes hard to discern before you install it and run it yourself. Some deficiencies cannot be overcome. This is why it is important to do a good process evaluation before you buy the equipment.

OPERATE AND MAINTAIN THE TREATMENT SYSTEM

After the treatment system is up and running, document its performance. Keep records of how much wastewater is treated, the amount of treatment chemicals and disposables (such as filters), and how much time is spent running and maintaining it. If significant, keep track of utility (electricity, water, air, gas, steam) expenses.

At minimum, sample and analyze the treated water as required by your discharge permit. You may want to collect and analyze more samples than required to ensure that your treatment system is functioning well. Prepare and submit the required records and reports to your pretreatment authority. And, of course, pay the required wastewater discharge fees.

Keep track of the amount of waste treatment residuals produced. Properly store and dispose of them.

Perform the required preventive maintenance to keep the treatment system performing well. Keep critical spare parts on hand and know where to obtain replacement parts quickly. If your treatment plant stops working you may have to stop generating wastewater until you fix it. In general, keep the plant area cleaned up. A cluttered, grimy work area creates a bad first impression.

8 WORKING WITH SUPPLIERS

INTRODUCTION

Most businesses rely on vendors and sometimes consultants to help them establish their system. Even if your reliance on vendors is minimal, you will be buying equipment and will want to ensure that the equipment operates reliably. You may decide to rely on a vendor or consultant to install your treatment system and you will want to make sure that you are working with someone who is competent. You will be "stuck" with whatever you install. The vendor or consultant will not have to live with the installation — *Caveat Emptor* (let the buyer beware).

Pollution control is big business. It is driven by regulations and public perception. Many vendors are trying to sell equipment. You need to sift through the many products and services offered, the hype, and the misinformation. Vendors and consultants can help you become aware of the pollution control options available to you. You must place the information into perspective and evaluate it. The purpose of this chapter is to suggest an approach (summarized in Table 26) to working with suppliers. The intent is to reduce the risks associated with relying on others to help design your system and with buying equipment.

Table 26 An Approach to Working with Suppliers

Determine if their experience is relevant to your situation
Investigate their reputation in the business
Define your problem before you call them
Establish what you expect them to provide
Obtain a cost estimate for their services or equipment
Ask for design, installation, and training proposals in writing
Ask them to test their equipment in your shop
Have the necessary site work done before equipment arrives for installation
Set the stage for site visits and training

WHAT IS A VENDOR?

"Vendor" is a term used for equipment and chemical suppliers. Some trades refer to them as jobbers or suppliers. A vendor may sell only equipment and chemicals, leaving you to do all of the design, building and training yourself. Some vendors will provide support and advice for the use of their chemicals and equipment, and will help you evaluate them. A vendor may do the entire job, from characterizing your wastewater through startup of the system and training your operators. Some vendors will even run and maintain a treatment system. No matter how much of the work you do yourself, in order to install your treatment system you will deal with vendors for something.

WHAT IS A CONSULTANT?

"Consultants" typically are not vendors. Their primary function is to provide you with information and help you install a successful system. A consultant typically does not sell equipment, but recommends what to buy. Some do sell equipment that they have designed or represent equipment manufacturers. A consultant could help prepare specifications and obtain quotations, and you may have a consultant manage the installation and startup of the system, prepare the manuals and permit application, provide training, and even run the system.

WORKING WITH SUPPLIERS

The same rules apply whether you are working with a vendor or a consultant. You need to ask similar questions of them and obtain similar information. You want to make sure the right decisions are made. Vendors and consultants will be referred to together here as suppliers, except where explicitly stated. Some things you should be aware of when working with suppliers are their experience level, reputation with other businesses and regulators, and their approach to your problem.

Experience Level

The knowledge and experience of suppliers vary widely. Some are just learning while others have been in the business a long time. You need to make an assessment of how much they really know. Ask about their background. Find out if they have worked with shops similar to yours. Have they previously installed treatment systems that work?

A company representative may only be getting started in learning about wastewater treatment. In this case you must determine if the company stands behind their representatives and products. Does the company employ people with the relevant expertise, and is he/she being supervised and trained? Will you have the benefit of the collective knowledge of the organization? Or will the company do a marketing job on you to sell their services and turn the project over to a rookie who is left on his/her own to learn by doing?

Is the supplier familiar with your type of operation or wastewater? If they have worked successfully with shops similar to yours, you have an indication that they know what they are doing. Some of the best suppliers are those who worked in the industry before they moved into sales or consulting. They have the perspective of an insider and may better understand the pressures and constraints with which you must deal.

A supplier may not have worked with your type of operation, but may be familiar with the processes and wastes that you generate. Someone who understands wastewater treatment can apply lessons learned from other jobs to yours regardless of whether the work was done in your industry. Pollution control regulations and technology are continually changing. You and your suppliers are forced to keep learning.

Is the company just branching out into the pollution control area to capitalize on the current environmental boom? Or is pollution control itself of keen interest? How committed are they to learning about pollution control? Do they understand how to apply wastewater treatment technology? Do they understand the regulations and what it takes to comply with them? They should be able to tell you what regulations apply to your situation. Are they prepared to spend their own time getting up to speed, or will you be paying for their education and mistakes?

The most important quality of a supplier is willingness to work with you to solve your problem. You must establish a good working relationship. Find a supplier who you can talk to, who "speaks your language". You have to learn enough about the wastewater treatment system and its requirements to be able to run and maintain it on your own when the supplier is gone. You need someone who will take the time to explain the process.

Reputation with Other Businesses and Regulators

Ask other shop managers about the wastewater treatment suppliers with whom they have dealt. Find out what they think of the service provided. You may identify suppliers who are good to work with and ones to avoid.

Ask the supplier for references and locations of equipment and visit the shops that are similar to yours. Talk with the plant managers and find out what testing and evaluation was done to install their systems. Ask what problems were encountered in installation and how the supplier responded to them. Determine if the level of technical support provided by the supplier was satisfactory.

Ask your local pretreatment authority for a list of suppliers who have installed systems that are meeting pretreatment requirements. This will provide another starting point for finding a supplier. At least you will know that the supplier has dealt with the agency and may therefore understand their permitting requirements.

The information that regulatory agency staff can provide is restricted. They cannot tell you who to avoid. Statements of that nature are too general. They try to respond objectively to specific questions. They will certainly review your pretreatment system plans and comment on the design process and whether it follows accepted practices. They may be able to tell you, based on their experience, what treatment technology is applicable to your wastewater and what is not. Ask them specific questions related to treatment technology rather than opinions of a particular supplier's product. Discuss technical aspects with them. Their comments can help you assess what you are being told by a supplier.

The Approach

Before You Contact the Suppliers

Define your problem before you contact the suppliers. Know the characteristics of your wastewater, your wastewater discharge limitations, and your budget. Prepare a plan, including the scope of work you expect the suppliers to provide and the schedule for implementation.

At the very least, you need to know the basic wastewater characteristics. Are you dealing with emulsified oily water, volatile organic or metal-bearing water, or a mix containing a variety of contaminants? Some suppliers specialize in certain types of wastewater treatment. Others can provide solutions for any type of wastewater. You need to know your wastewater discharge limitations. Suppliers should be able to give an indication as to whether their equipment can produce wastewater that meets the limitations. Screen out those who cannot work with your wastewater.

When You Contact the Suppliers

Tell the suppliers what you expect them to provide. Make sure it is clear up front that they understand their role. Ask them for a cost estimate for their services and equipment. Ask for design, installation, and training proposals in writing. If you do not know what you are doing, spend the money to find out. You will benefit in the long run.

A consultant will charge for any work done to find out what they need to know to design a system and for preparing the necessary documentation. An equipment or chemical supplier typically includes the cost of coming out to your shop, getting information, and preparing a design in the cost of what they sell. Screen out the suppliers you cannot afford or who cannot provide the services you need.

The supplier should ask for pertinent information. In fact, he/she should be quite demanding in asking how things really operate at your company and what you expect out of a treatment system. Be wary of a supplier who tells you they have just what you need before time is spent in your shop.

Consider what you are getting for the money you spend for the design and equipment. Quality technical assistance and equipment cost more than second-rate products. You may have to spend more per hour to hire a consultant who has extensive experience, but the consultant may be able to work more effectively and faster than an inexperienced consultant. You may spend more for a treatment system that requires less attention and maintenance, but make up the cost difference over the life of the equipment. If you decide to go with the cheapest vendor or consultant, make sure that they are reputable. If you buy equipment at a bargain, make sure that at least it can produce acceptable treated water.

Some suppliers offer only one type of treatment equipment. You must decide if the equipment is right for your shop. The supplier may let his/her enthusiasm for the equipment override any objectivity and lead to misapplication of the technology. Do not rely on suppliers' claims. Ensure that the equipment will work for you. Have them test it out.

Use suppliers as a resource. A good supplier is interested in providing information which will help you make decisions. Obtain all the information and supporting data possible from them. Some equipment and chemical suppliers will provide much good objective information. They may provide literature and articles that explain the technology and compare it to other options, well-documented test results, and data from operating systems. The best suppliers will point out the disadvantages as well as the advantages of their technology. They will tell you honestly they do not think their product will meet your needs and may suggest more viable options.

Some suppliers are merely self-serving. They want to sell you their products whether or not they are the best choice for you. They may not know any alternative because they do not have the experience needed to compare alternatives and select the best one. They will feed you information that makes their product or service sound fantastic. Beware of overly enthusiastic suppliers who make outrageous claims.

It is your job to assess the technology; the supplier's job is to bring it to your attention. When you find a technology that you like, visit a site to see it working. The supplier should be able to provide you with such sites. The shops you visit should have wastewater that is similar to yours. The shop owners should be willing to talk about how well the system is working for them and what problems they have experienced using it.

When the Suppliers Visit

Ask the supplier to test the equipment in your shop. Some suppliers will provide pilot test equipment. The supplier should be at least willing to conduct bench tests on your wastewater to help ensure that their process will work. If the supplier is unwilling to test your wastewater, then proceed with caution. It may be that he/she understands your wastewater well enough to treat it without testing it. Just be sure that good solid evidence is presented to back up that claim.

Define what support you expect from the supplier during installation and after the system is installed and initiated. Think about training, maintenance, and technical support. Make sure that it will be provided. It is best to have such assurances in writing to reduce the chance of misunderstandings. Do not rely on vague statements such as, "I'll probably be available if you need help." Ask for what you want or you may not get it. You cannot assume that the supplier will provide installation, training, or follow-up support.

When the supplier is at your site, be ready to accommodate him/her as time is limited. Have requested information ready before he/she arrives. If wastewater samples are needed, make sure that you will have wastewater available. If the supplier is installing the equipment, make sure you provide whatever is needed. All the necessary site work, such as installation of utilities and equipment pads and berms, should be in place before the supplier arrives. Any special tools, such as forklifts and cranes, should be onsite. Do not wait until the installer is at the door to think about this.

The supplier may provide training with equipment installation. Do not confuse learning how to run the equipment with learning how to treat wastewater. Learn all you can about the treatment process before you learn how to run the equipment. Read the instruction manual. Do not wait until the equipment is installed to start learning about it. Optimize the use of the training time. You must have some background to understand what you are being taught. This is especially important if you are under a strict compliance schedule and must meet discharge limitations as soon as the equipment is installed. Consider videotaping the training for future reference.

Set the stage for site visits and training. Tell your staff the purpose of the supplier's visit. Make sure they understand what is required of them. Explain that the supplier needs their attention to do a good job. Be aware of the time required and make it a priority to provide it. Minimize distractions by scheduling work around the site visit or training time. Allow extra training time if the treatment process or equipment is unfamiliar to shop workers or if your shop's processes are initially unfamiliar to the supplier.

9 ANNOTATED BIBLIOGRAPHY

INTRODUCTION

The foregoing chapters provide a outline and description of the treatment process, and alert you to things you need to consider when implementing a treatment system, but they do not cover all aspects of wastewater treatment. The subject is too vast to be encompassed by a single book. A number of periodicals and reference books pertain to wastewater treatment which you might find useful. This chapter provides an introduction to some of them. The author owns most of the books reviewed here and refers to them often.

It is unlikely that any one book will answer all of your questions and put things in terms you can understand and apply. Pick several books which together cover what you need to know.

To find books, visit your local library and bookstores. Keep an eye out for used books and publisher remainder sales. Former editions of books still in print make useful references and can be found at greatly reduced prices. Ask what books other businesses have found helpful. Contact your trade association for references. Read book reviews in periodicals. Visit publisher booths at trade shows and conferences.

Publishers mail book descriptions to those on the mailing lists of the periodicals listed below. Suppliers also use the periodicals to reach businesses through their mailings. Once you start subscribing to the periodicals, your name will be put on their lists. Ask publishers for catalogs. A number of publishers will let you review a book for 15 to 30 days before you buy it.

On-line sources are becoming more accessible and can provide information applicable to wastewater treatment system development. Available information includes: MSDSs, periodical indexes, hazardous material toxicity data, federal regulations and bills, descriptions of research undertaken at various institutions, and publishers' book listings with descriptions and

Table 27 List of the References Reviewed in Chapter 9

Chemistry and hazard references
Dangerous Properties of Industrial Materials
Standard Methods for the Examination of Water and Wastewater
Test Methods for Evaluating Solid Waste (EPA/SW-846)
The Condensed Chemical Dictionary
Information source references
ACCESS EPA
American Chemical Society Environmental Buyers' Guide
Chemical Engineering Equipment Buyer's Guide
Pollution Engineering Yellow Pages
Pollution Equipment News Buyer's Guide
Water Environment Federation Buyer's Guide & Yearbook
Regulatory references
Guide to Environmental Laws
Guide to State Environmental Programs
Pollution Control Engineer's Handbook
Regulatory Requirements for Hazardous Materials
Wastewater treatment technology references
Biotreatment of Industrial and Hazardous Waste
Carbon Adsorption Isotherms for Toxic Organics
CERCLA Site Discharges to POTWs, Guidance Manual
Chemical Engineering
Electroplating Wastewater Pollution Control Technology
Groundwater Treatment Technology
Hazardous Waste Management Engineering
Hazardous Waste Treatment Processes, Manual of Practice No. FD-18
Industrial Waste Treatment, A Field Study Training Program
Industrial Wastewater Treatment Technology
Ion Exchange Training Manual
Maritime Industrial Waste Project
Pollution Engineering
Pollution Equipment News
Pretreatment of Industrial Wastes, Manual of Practice No. FD-3
Standard Handbook of Environmental Engineering
Standard Handbook of Hazardous Waste Treatment and Disposal
The Nalco Water Handbook
The Operation and Maintenance of Surface Finishing Wastewater Treatment Systems
Treatment of Metal Waste Streams, A Field Study Training Program
Ultrafiltration Handbook
Wastewater Engineering, Treatment, Disposal and Reuse
Wastewater Treatment by Ion Exchange
Water Treatment Plant Design

ordering information. You can also find news groups where you can post questions and get answers. The author has established a Virtual Library — Industrial Wastewater Engineering home page on the Internet. You can access it via URL: http://www.halcyon.com/wastewater/. Rather than describe in detail what information is available there, I suggest you take a look.

Table 27 contains a list of the references reviewed in this chapter. They are arranged by subject and then alphabetically by title. Publisher information is provided in each review. Figure 13 shows some of the reviewed books.

Figure 13 Some of the books and periodicals reviewed in Chapter 9.

CHEMISTRY AND HAZARD REFERENCES

Dangerous Properties of Industrial Materials, 9th Edition
N. Irving Sax and Richard J. Lewis, Sr.
Van Nostrand Reinhold, 115 Fifth Avenue, New York, NY 10003
1993 3 volumes Write for information.

The objective of this multivolume work is to promote safety by providing the most up-to-date hazard information available. The books contain references to some 20,000 chemicals. CAS Registry, NIOSH, RTECS, and DOT Guide numbers are included where these have been assigned. Entries are listed alphabetically and are crossed indexed by CAS Number and synonyms. This gives the reader a powerful tool to find information about a compound listed on an MSDS or product label where a common name or trade name is often used. And the CAS Number can be used to search other databases for more information.

Each material is assigned a hazard rating (low, medium, or high) that indicates the level of toxicity or fire, explosive, or reactivity hazard. A "D" rating is assigned where the available data are insufficient to provide a relative rating. Entries include molecular weights, formulas, flash points, explosion limits, and other pertinent physical data. Clinical data on humans and experimental animals, including LD_{50} and LC_{50} are included with references to the primary literature. This information can help when planning for or responding to an emergency situation. International Agency for Research on Cancer (IARC) Review conclusions and the U.S. National Toxicology Program cancer testing results are provided. The IARC reviews are conducted under vigorous scientific peer review conditions and their conclusions are based on published, reviewed literature.

References to the EPAs list of chemicals of interest are provided. Notations refer to the EPA Extremely Hazardous Substances List, the EPA TSCA Inventory, and the Community Right-to-Know List. OSHA, ACGIH, and NIOSH worker exposure limitations and DOT classifications are provided.

Table 28 Summary of the Contents of Standard Methods

Part	Contents
1000	Information regarding the proper execution of the procedures given in the manual
2000	Physical and aggregate properties
3000	Metals analysis
4000	Inorganic nonmetallic constituents
5000	Aggregate organic constituents
6000	Individual organic compounds
7000	Radioactivity
8000	Toxicity
9000	Microbiological examination
10000	Biological examination

Toxic and Hazard Reviews are provided, which summarize the information about the material.

This is an excellent reference. It provides a wealth of information useful to a pollution control manager. The only drawback to having your own copy is the price. However, you may be able to find a previous edition for $30 to $100. The author purchased the 5th edition for $100 in 1989, and the 7th edition for $30 in 1993. An earlier edition, while not completely up to date, still contains much useful data.

Standard Methods for the Examination of Water and Wastewater, 18th Edition
Prepared by: American Public Health Association (APHA), American Water Works Association, and Water Environment Federation
Publication Office: APHA, 1015 15th St., Washington, DC 20005
1992 1100 pages $160.00

The methods presented in this book are the best-available, generally accepted procedures for the analysis of water, wastewater, and related materials. They represent the recommendations of specialists, ratified by a large number of analysts and others of more general expertise, and as such are consensus standards, offering a valid and recognized basis for control and evaluation. Each procedure includes recommendations on methods selection, sampling and sample storage, and interferences. Table 28 presents a summary of the book's contents.

This book is recognized as a standard reference for laboratory analysis of water and wastewater. The methods are accepted by most state and local authorities for the analysis of wastewater to determine if it meets their discharge limitations. You will find a copy of it in most analytical laboratories. *It includes information that a pollution control manager will find useful, especially if he/she needs to know the details of chemical analysis or is just interested in the chemistry. It is written from a chemist's perspective and most of the methods require special equipment.*

Test Methods for Evaluating Solid Waste, 3rd Edition (EPA/SW-846)
Volume 1A through 1C, and Volume 2, *Field Manual Physical/Chemical Methods*
Environmental Protection Agency, Office of Solid Waste and Emergency Response
Available from: National Technical Information Service, Springfield, VA 22161
1986 1748 pages

This manual provides test procedures which may be used to evaluate those properties of a solid waste which determine whether the waste is a hazardous waste within the definition of Section 3001 of RCRA. These methods are approved for obtaining data to satisfy the requirement of 40 CFR Part 261, Identification and Listing of Hazardous Waste. Volume 1A deals with quality control, selection of appropriate test methods, and analytical methods for metals. Volume 1B consists of methods for organic compounds. Volume 1C includes a variety of test methods for miscellaneous analytes and properties. Volume 2 deals with sample acquisition, and includes quality control, sampling plan design and implementation, and field sampling methods.

These methods are cited in state and local regulations and are accepted by many authorities for analyzing wastewater to determine if it meets their discharge limitations. You will not be able to follow the procedures unless you are a chemist and have the necessary equipment. *However, this is one of the standard analytical references, and it is found in virtually all of the laboratories that perform chemical analysis for regulatory purposes.*

The Condensed Chemical Dictionary, 10th Edition
Revised by Gessner G. Hawley
Van Nostrand Reinhold, 115 Fifth Avenue, New York, NY 10003
1993 1000 pages

The Condensed Chemical Dictionary gives summaries of pertinent facts about thousands of chemicals and chemical phenomena. Three types of information are presented, including: technical descriptions of chemicals, raw materials and processes; expanded definitions of chemicals, phenomena and terminology; and descriptions or identification of trademarked products used in the chemical industries.

The following information is provided, when available, for each chemical substance listed: chemical name, synonyms, chemical formula, properties, source, derivation, available grades, available unit container types, hazards, uses, and shipping regulations. The terminology descriptions include group definitions (acid, base), chemical and physicochemical phenomena (flocculation, coagulation, sedimentation), functional names (antifreeze, lubricant), and chemical processes. Trademark information relates the trademark name to the common chemical name.

The Condensed Chemical Dictionary is an excellent desktop reference book. It provides quick answers to common questions regarding chemical terminology. It is a practical guide which can help take the mystery out of MSDSs and product labels.

REGULATORY REFERENCES

Guide to Environmental Laws — From Premanufacture to Disposal
J.A. Stimson, J.J. Kimmel, and S.T. Rollin
The Bureau of National Affairs, Inc.
1250 23rd St., N.W., Washington, D.C. 20037-1165
1993 338 pages $48.00

This guide explains the functions and interrelationships of the environmental laws, regulations, and enforcement strategies that affect every phase of the manufacturing process. The laws are grouped into six sections that coincide with major phases of the manufacturing process: premanufacture controls, Right-to-Know requirements, manufacturing controls, transport and identification, storage, disposal, and cleanup. *The book is an overview of federal legislation which may help you to understand the regulations to which you are subject, provided your state and local authorities have not promulgated more stringent regulations.*

Guide to State Environmental Programs, 2nd Edition
Deborah Hitchcock Jessup
The Bureau of National Affairs, Inc.
1250 23rd St., N.W., Washington, D.C. 20037-1165
1990 700 pages $62.00

This guide presents descriptions of each state's program. For each state it gives a statement of the organizational scheme, with applicable agency names, agency addresses and telephone numbers, its unique characteristics, and its priorities. The most effective first contact is noted. The state-designated Emergency Response Commission address and telephone number are given. Summaries of the major programs covering air and water pollution, wetlands protection, water use, and solid, infectious, and hazardous waste management are provided. It also includes as appendices directories of the federal agencies most likely to play a part in state permitting activities, as well as state, regional, and local agencies when too numerous for inclusion in the text, but they carry out important roles. *This guidebook can help you to determine who you should be talking to and can provide some perspective on what businesses in other states are faced with. It may be especially useful to you if you do business in several states and need to find program contacts.*

Pollution Control Engineer's Handbook
Compiled by Edward J. Shields
Pudvan Publishing Co., Inc., 1935 Shermer Road, Northbrook, IL 60062
1985 144 pages $19.95

This handy book is a collection of data and information related to air, water, groundwater, and solid waste environmental concerns. The regulatory references are dated, but provide a good basis for understanding the myriad regulations that are faced. You can call the listed state and EPA regional regulatory offices for the latest versions and interpretations of the regulations. A list of acronyms and glossary can help you decipher the dry regulatory tomes. Information is provided on the BOD of organic compounds, activated carbon absorption efficiencies, neutralization agents, trace metal content of natural soils, biodegrability of organic compounds, and unit conversions. *This practical handbook contains a variety of material of interest to technical types with references that can lead you off for more. An expanded second edition would be welcome.*

Regulatory Requirements for Hazardous Materials
Somendu B. Majumdar
McGraw-Hill, 1221 Avenue of the Americas, New York, NY 10020
1993 524 pages $70.00

This book presents a comprehensive review of various environmental statutes and regulations. The object is to bridge the gap between professionals in the scientific, legal, and cooperate communities to help develop an understanding of hazardous material regulations. It includes an overview of federal statutes, carcinogens, hazardous waste toxicology, air toxics under the Clean Air Act, hazardous waste under RCRA, hazardous substances under CERCLA, emergency planning and Right-to-Know under SARA Title III, pesticides under FIFRA, toxic substances under TSCA, spill reporting, radioactive waste, remediation technology, enforcement, liability, compliance, negotiation, and settlement. It summarizes the important federal regulations. *This book can help you understand the technical and legal issues, from a federal regulatory perspective, which surround the use, handling, and disposal of hazardous materials. Your state and local authorities may have promulgated more stringent regulations.*

WASTEWATER TREATMENT TECHNOLOGY REFERENCES

Biotreatment of Industrial and Hazardous Waste
Morris A. Levin and Michael A. Gealt, Editors
McGraw-Hill, 1221 Avenue of the Americas, New York, NY 10020
1993 331 pages $60.00

This book is an in-depth guide to evaluating, selecting, and applying microbial techniques to biologically treat industrial and hazardous waste. It includes contributions by 22 scientists and engineers. The basic principles of biological treatment are explained, including the effects of site constraints, the selection of microorganisms, and the type of bioreactors used. The laws and policies regulating bioremediation of hazardous wastes are explained. Modeling of biodegradation and the fate of the organic compounds is examined. Bench testing and field implementation examples are provided.

This technical reference book includes over 800 literature and regulatory citations. It is not an introductory text. You will need some background in biological treatment to make full use of it. It could provide a satisfying challenge for the neophyte and the opportunity to get started in the right direction.

Carbon Adsorption Isotherms for Toxic Organics
Richard A. Dobbs and Jesse M. Cohen
U.S. Environmental Protection Agency, Wastewater Research Division, Cincinnati, OH 45268
1980 332 pages

This reference presents an experimental protocol for measuring activated carbon isotherms and applies it to a range of organic compounds. Compounds from the OSHA list of regulated carcinogens and the EPA priority pollutant list are included. Adsorption capacities range from 11,300 mg/g for bis (2-ethylhexyl) phthalate, which is very amenable to carbon treatment to 0.000068 mg/g for *N*-dimethylnitrosamine, which is not treatable by carbon adsorption. Thirty-two literature citations are provided for further research. *This book is an excellent source of carbon adsorption capacity data that you can use to evaluate the potential effectiveness of carbon adsorption for the removal of organics from your wastewater.*

CERCLA Site Discharges to POTWs, Guidance Manual
U.S. Environmental Protection Agency, Industrial Technology Division
401 M Street, S.W., Washington, DC 20460
Available from the National Technical Information Service, Springfield, VA 22161
1990 203 pages

The stated purpose of this guidance manual is to provide feasibility study writers, USEPA Remedial Project Managers, state officials, and POTW personnel with the current regulatory framework and technical and administrative guidance that is necessary to evaluate the remedial alternative of discharging wastes from CERCLA sites to POTWs. The regulatory basis of CERCLA site discharge is interesting, but only indirectly applies to the discharge of industrial waste.

This book is included as a reference because it includes an approach to characterizing wastewater, evaluating pretreatment requirements, and identifying and screening pretreatment alternatives. It is also useful for POTW staff because it provides an approach to establishing local limits for hazardous wastewater contaminants. It includes some 40 references for further study. *The book is an example of the type of EPA publication that you can use to gather ideas to develop your pretreatment system.*

Chemical Engineering
Richard J. Zanetti, Editor-in-Chief
McGraw-Hill, 1221 Avenue of the Americas, New York, NY 10020
published monthly 150–200 pages $30/year subscription

Chemical Engineering is a trade publication directed at those in the chemical process industry. Each issue contains general commentary, news articles, new product and service information, process engineering and operation and maintenance articles, a column called "You & Your Job", and information on the use of computers in the industry.

Although much of the information is most applicable to large industries, the pollution control manager at a smaller facility will find interesting articles and useful information. The product information alone is worth the cost of the subscription and you may find something you need in the new/used equipment classified section. The annual Chemical Engineering Buyers' Guide, reviewed elsewhere, is included with the subscription.

Electroplating Wastewater Pollution Control Technology
George C. Cushnie, Jr.
Noyes Publications, Mill Road, Park Ridge, NJ 07656
1985 239 pages $36.00 (out of print)

This book presents more than 27 pollution control technologies applicable to electroplating wastewater. It includes a section on in-plant process changes that may help reduce overall chemical and water usage and consequent water treatment and disposal costs. Chemical precipitation, oxidation and reduction, ion exchange, evaporation, membrane technology, sludge dewatering treatment chemistry, flow diagrams, and equipment diagrams are included.

Electroplating wastewater typically contains soluble metals, and wastewater from other industries may contain more particulate material, but the described treatment processes are also applicable to wastewater generated by other industries. As the author notes, the selection of any of the described technologies should be done only after rigorous identification of site requirements. *Process recipes and equipment requirements are covered in enough depth to provide you with ideas for your situation whether or not you are a metal finisher.*

Groundwater Treatment Technology, 2nd Edition
Evan K. Nyer
Van Nostrand Reinhold, 115 Fifth Avenue, New York, NY 10003
1992 320 pages $51.95

This book explains how to apply technology and engineering methods to groundwater decontamination. Comparisons are made between wastewater and groundwater clean up. Physical/chemical methods for organic contaminant removal, including air stripping and carbon adsorption, are presented. Biological methods for organic contaminant removal, inorganic contaminant treatment methods, and *in situ* organic removal methods are included. For each method, information to design and operate effective treatment systems is provided. Case studies examining implementation strategies and literature references are provided.

If you are faced with a groundwater remediation project, this book can help you install an effective treatment system. *If you are treating industrial wastewater, especially organic contaminated wastewater, this book can give you suggestions to help you through the design and implementation process. The explanations of how the treatment of groundwater and wastewater differ will help you understand the design process.*

Hazardous Waste Management Engineering
Edward J. Martin and James H. Johnson, Jr., Editors
Van Nostrand Reinhold, 115 Fifth Avenue, New York, NY 10003
1987 520 pages $99.95

This reference presents basic technical information for the evaluation of competing hazardous waste management technologies. It gives a general overview of hazardous waste regulations and risk assessment. It includes chapters on hazardous waste incineration, storage, land disposal, and facility siting. Literature references are provided.

The chapter on chemical, physical, and biological treatment of hazardous waste includes a treatability summary table with references for 505 chemical compounds. For each compound a summary statement is provided with information on treatability with biological, precipitation, reverse osmosis, ultrafiltration, stripping, solvent extraction, carbon adsorption, resin adsorption, and miscellaneous adsorbent technology. Five concise examples of treatment applications are given. The chapter on landfill leachate management describes leachate treatment design methodology and gives brief treatment system examples. *The principles applicable to leachate treatment system design also can be applied to industrial wastewater treatment system design and may help you to design your treatment system. The book has a broad scope but contains some good tidbits of information.*

Hazardous Waste Treatment Processes, Including Environmental Audits and Waste Reduction, Manual of Practice No. FD-18
Task Force on Hazardous Waste Treatment
William J. Librizzi and Catherine N. Lowery, Co-Chairpersons
Water Pollution Control Federation, 601 Wythe Street,
Alexandria, VA 22314-1994
1990 332 pages paper $60.00

The purpose of this manual is to provide a description of the technical and regulatory approaches to hazardous waste treatment. The most widely applied technologies, including biological, physical, chemical and thermal processes are explained. It briefly describes waste characterization and treatability and pilot testing. It also includes guidelines for conducting environmental audits and approaches to waste reduction. It is intended for use and reference in planning, executing, and continuing a program of hazardous waste treatment. *The information provided in this manual is applicable to industrial wastewater treatment and it contains other information that a pollution control manager will find helpful.*

Industrial Waste Treatment, A Field Study Training Program
Kenneth D. Kerri, Project Director
Prepared by California State University, Sacramento, for the EPA
Available from Ken Kerri, California State University,
Sacramento, 6000 J Street, Sacramento, CA 95819-2694
1989 611 pages $20.00

This operator training manual covers the importance and responsibilities of an industrial wastewater treatment plant operator. It was originally developed as a home study course and includes an objective test at the end of each chapter. It can be used for self-study or as the basis of a training program.

It includes the activated sludge process, physical/chemical treatment, instrumentation, industrial waste monitoring, industrial waste treatment processes, and maintenance. The industrial monitoring chapter covers sampling objectives, types of samples and sampling equipment, sample preservation, and sampling strategies. The industrial waste treatment chapter includes explanations of neutralization, coagulation and precipitation, and carbon adsorption processes. The maintenance chapter describes how to select and maintain portable pumps, pipes, valves, and fittings, and safely operate and maintain auxiliary equipment. A final exam and glossary are also included. *This book contains a lot of information. You will probably find something useful and it is a bargain.*

Industrial Wastewater Treatment Technology, 2nd Edition
James W. Patterson
Butterworths Publishers, 80 Montvale Avenue, Stoneham, MA 02180
1985 467 pages $72.95 (out of print 12/93)

This book focuses on treatment technology that has been proven in full-scale operation. Its contents are arranged by contaminant and include chapters on aluminum, arsenic, barium, cadmium, chromium, copper, cyanide, fluoride, iron, lead, manganese, mercury, nickel, nitrogen, oil and grease, toxic organics, pH control, phenol, selenium, silver, total dissolved solids, and zinc. Sources and typical levels of the pollutants are identified and described. Available treatment technology is described with respect to how each technology operates, its limitations, what levels of treatment have been accomplished, and relative costs of each technique. A concise summary of the major technologies and attainable effluent levels for each contaminant is provided.

This book is an excellent source of information on treatment technologies that are in common use. It includes extensive information about chemical precipitation techniques for metals and oil and grease. Data are given for the biological removal of the 112 organic priority pollutants and activated carbon adsorption of 43 of the compounds. Numerous references are provided. *If you are interested in chemical precipitation technology, this is a must-have book.*

Ion Exchange Training Manual
George P. Simon
Van Nostrand Reinhold, 115 Fifth Avenue, New York, NY 10003
1991 230 pages $49.95

This book reviews the development of ion exchange materials. It describes different types of ion exchangers and the techniques used to produce them. Ion exchange properties related to ionization and hydration are explored. Applications are described for water treatment, industrial applications, and pollution control, including deionization, demineralization and electrodialysis. Limitations and problems involved with the technology are described. Sample design calculations are provided for two- and three-bed systems to help you analyze incoming water, select necessary types of pretreatment (prior to ion exchange), evaluate types of systems and resins, determine regenerant levels and capacities, determine the length of service run and flow rate, and make final design calculations. A list of ion exchange suppliers, laboratory test procedures, glossary of terms, and suggested reading list are provided. *This book is a good starting point if you are serious about ion exchange.*

Maritime Industrial Waste Project, Reduction of Toxicant Pollution from the Maritime Industry in Puget Sound
Metro Water Pollution Control Department

Industrial Waste Section, 130 Nickerson Street, Suite 200,
Seattle, WA 98109-1658
1992 150 pages limited availability

This reference reports the results of a project conducted to characterize the wastewater generated from hull washing operations and to provide pretreatment system design criteria and waste disposal guidelines. It describes the selected boatyards and shipyards, the wastewater characterization procedure, and the treatment technologies tested. It includes the results of pilot testing conducted at maritime facilities and provides conceptual designs and cost estimates. The report is acceptable to the pretreatment authority as the design basis for a maritime hull wash water treatment facility, provided that one of the technologies demonstrated to produce acceptable treated wastewater is used.

This is an example of the type of documentation, prepared by a pretreatment authority, that can be very helpful to you in your undertaking. It is the product of a year-long project undertaken to provide practical pretreatment system guidance to the maritime industry. It describes treatment system design from waste characterization, through process screening, pilot testing, and conceptual design.

Pollution Engineering
Diane Pirocanac, Chief Editor
Cahners Publishing Company, Des Planes, IL
(send all subscription mail to: Pollution Engineering,
44 Cook St., Denver, CO 80206-5800; published monthly,
100–150 pages, no charge to qualified subscribers, plus annual
Pollution Engineering Directory published in October)

This is a trade publication directed toward pollution control managers. It contains practical articles to help you comply effectively with the regulations. Each issue offers regulatory, industrial, and technological news, a brief regulations review, casebook descriptions of installed pollution control processes, a pollution prevention case study, a legal outlook on a chosen topic, and numerous product advertisements and descriptions of available product literature. An annual Pollution Engineering Yellow Pages directory to suppliers (reviewed elsewhere) is included with a subscription. *If you are responsible for pollution control, subscribe to this one. It is great!*

Pollution Equipment News
Rimbach Publishing Company Inc., 8650 Babcock Blvd.,
Pittsburgh, PA 15237-5821(published monthly, 75–100 pages, no charge

to qualified subscribers, plus annual *Pollution Equipment News Buyer's Guide* published in November)

This is a trade publication directed toward pollution control managers. It contains hundreds of equipment advertisements, short product descriptions, descriptive articles written by suppliers, short articles on various topics, handy equipment selection charts, and an equipment section. An annual *Pollution Equipment News Buyer's Guide* (reviewed elsewhere) is included with a subscription. *If you are responsible for pollution control, subscribe to this one.*

Pretreatment of Industrial Wastes, Manual of Practice No. FD-3
Task Force on Pretreatment of Industrial Wastes
Elin Eysenbach, Chair
Water Environment Federation, 601 Wythe Street,
Alexandria, VA 22314-1994
1994 252 pages $95.00

This practical manual provides design and operational information about industrial waste pretreatment systems preceding municipal wastewater treatment plants. The objective of the manual is to provide guidance in the selection of designs and processes. It is organized by waste characteristics. Each chapter describes the potential effects of the specific waste characteristics on the operation and efficiency of a POTW. Alternative pretreatment design concepts and operational considerations are described and advantages and disadvantages compared.

This manual presents descriptions of various treatment technologies and waste management practices and references are provided. The examples tend to show large equipment; however, the processes shown can also be done on a small scale. In-plant controls to minimize the waste requiring treatment including waste reduction, water conservation, recycling, and process modifications are emphasized. Sound familiar? Some 15 years after the manual was first written, it is called pollution prevention. Good housekeeping has been well recognized for years as sound waste management practice. *This book will help you understand how industrial wastewater can impact municipal collection and treatment and what can be done to minimize that impact.*

Standard Handbook of Environmental Engineering
Robert A. Corbitt
McGraw-Hill, 1221 Avenue of the Americas, New York, NY 10020
1990 1360 pages $105.50

This reference on the principles and practices of environmental engineering was written by a team of engineers and scientists. Chapters are provided on legislation, air quality control, water supply, wastewater disposal, solid waste management, storm water management, hazardous waste management, and environmental assessment. The handbook has concise descriptions of a wide variety of subjects and includes many tables and figures. Selected and suggested references are provided to assist additional research.

Much of the information pertains to engineering projects that a small business would not normally undertake, such as groundwater well development, municipal wastewater collection and treatment, urban storm water management, municipal solid waste disposal, and hazardous waste disposal. However, a small business owner may find it useful to avail himself/herself of another perspective. The summary of environmental regulations is useful and includes descriptions of the major air and water quality and hazardous waste regulations. The descriptions of waste treatment methods and residual management options can help you develop your alternatives evaluation plan. *This is a decent general reference which provides a good overview.*

Standard Handbook of Hazardous Waste Treatment and Disposal
Harry M. Freeman, Editor in Chief
McGraw-Hill, 1221 Avenue of the Americas, New York, NY 10020
1989 1120 pages $105.50

This handbook gives a broad overview of current and proposed legislation for hazardous waste treatment and disposal, alternative waste minimization and recycling methods, physical and chemical treatment of waste streams, thermal treatment, and biological treatment. It contains 74 sections, of which 8 address the various types and categories of hazardous waste, 42 describe the various technologies, and 11 discuss land disposal and remedial action. The other sections include sampling and analysis techniques, and an overview of regulations and engineering considerations for hazardous waste management. A reference listing can lead you to other useful literature.

The sections on waste minimization and waste exchanges can help you get started thinking about waste reduction and recycling. The treatment technology descriptions include a process summary, principles of operation, limitations of the process, and process selection, design, and operational considerations. *Some of the technologies can be used for small business wastewater treatment. The descriptions of hazardous disposal facilities will give you insight into the reasons disposal is so costly.*

The Nalco Water Handbook, 2nd Edition
Nalco Chemical Company
Frank N. Kemmer, Editor
McGraw-Hill, 1221 Avenue of the Americas, New York, NY 10020
1989 1056 pages $83.50

This handbook covers incoming water supply and treatment technologies. It focuses on the conditioning and use of water in industrial processes. It includes descriptions of water sources, chemistry, contaminants, treatment, and disposal. The sections on coagulation and flocculation, solids/liquid separation, precipitation, emulsion breaking, and neutralization offer a good basic explanation of the physical and chemical aspects of these processes, which can be used to condition incoming waste or to treat wastewater. The sections on cooling water and boiler water treatment will be useful to plant managers responsible for the upkeep of those facilities. *This book can help provide a grounding in water technology. It has a strong chemistry focus.*

The Operation and Maintenance of Surface Finishing Wastewater Treatment Systems
Clarence H. Roy
American Electroplaters and Surface Finishers Society, 12644 Research Parkway, Orlando, FL 32826-3298
1988 199 pages trade $60.00, paper $40.00

This practical book focuses on plating wastewater treatment. It includes chapters on sources and discharge limitations, water and chemical conservation and recovery, chemical principles and concepts, treatment process instrumentation and control, operation of treatment processes, sludge dewatering, treatment system maintenance, and safety and health. *The descriptions are clear and directed at the problems faced by a treatment plant operator. The information will be useful to the operator of any metal treatment facility.*

Treatment of Metal Waste Streams, A Field Study Training Program
Kenneth D. Kerri, Project Director
Prepared by California State University, Sacramento, for the EPA
Available from Ken Kerri, California State University, Sacramento, 6000 J Street, Sacramento, CA 95819-2694
1990 130 pages 2nd edition, 1993, $10.00

This operator training manual was developed as a home study course and includes an objective test at the end of each chapter. It can be used for self-study or as the basis of a training program. It focuses on metal-finishing wastewater, but provides information useful to the treatment of other metal-bearing wastewater.

It covers how to operate, troubleshoot, and maintain facilities which neutralize acidic and basic wastewater, treat wastewater-containing metals, destroy cyanide, and treat complexed metal wastewater. Excellent descriptions of pH and ORP measurement are provided. A final exam and glossary are also included. *This book contains a great deal of information. You will probably find something useful in it and it's a bargain.*

Ultrafiltration Handbook
Munir Cheryan, Ph.D.
Technomic Publishing Co., Inc.
851 New Holland Avenue, Lancaster, PA 17604
1986 369 pages $65.00

This textbook presents an introduction to the definitions and classification of membrane processes, membrane chemistry and materials, membrane properties, models, and equipment. Factors influencing fouling and process design are described. A variety of field applications covering several broad industrial categories are given. The biotechnology and food processing industries are of central concern, but oily wastewater and electrocoat paint recovery are also included. Lists of membrane manufacturers and suppliers, references, books, and journals, and a glossary are included. *This book provides a overview of ultrafiltration that those with a technical background will appreciate.*

Wastewater Engineering, Treatment, Disposal and Reuse, 3rd Edition
Metcalf & Eddy (Revised by George Tchobanoglous and Franklin L. Burton)
McGraw-Hill, 1221 Avenue of the Americas, New York, NY 10020
1991 1334 pages write for trade information

This classic textbook was revised to keep pace with changing technology and legislation. It addresses the treatment of municipal wastewater and emphasizes biological treatment. It covers wastewater characterization, process selection, unit processes, and system design. The removal of toxic compounds and refractory organics is barely touched on, however. *The book will be useful for those considering installing a biological treatment system for BOD-containing wastewater who want to understand the details.*

Wastewater Treatment by Ion Exchange
B.A. Bolto and L. Pawlowski
E. & F.N. Spon, 29 West 35th Street, New York, NY 10001
1987 262 pages $69.95

This book describes the technical aspects of ion exchange technology and its application to industrial wastewater treatment. It deals with the particular problems of industrial wastewater, such as high turbidity, fouling and precipitation potential, and oxidizing properties. Flowsheets present a guide to applications in the recovery and reuse of water and chemicals. It focuses on methods in use in industry or at the pilot stage. It provides a list of resin suppliers and commercial resins. *If you are contemplating ion exchange, and can handle a technical reference, this book is recommended.*

Water Treatment Plant Design, 2nd Edition
American Society of Civil Engineers & American Water Works Association

McGraw-Hill, 1221 Avenue of the Americas, New York, NY 10020
1990 598 pages $65.00

This technical book addresses the design of drinking water treatment facilities. It includes a chapter on chemicals and chemical handling which describes the properties and characteristics of 50 chemicals used in water treatment. Recommendations are included for the storage, handling, feeding, and safe use of these chemicals as required by designers and supervisors of water treatment operations. A chapter describes a number of design considerations (other than the treatment process) that are important to any treatment plant design. The trace organics and special water treatment processes chapters describe processes for the removal of organics and metals which are applicable to industrial wastewater treatment. *Although the primary focus is not industrial wastewater, this book presents information useful for the design of any wastewater treatment plant.*

INFORMATION SOURCE REFERENCES

ACCESS EPA
U.S. Environmental Protection Agency, Information Access Branch
401 M Street, S.W., Washington, DC 20460
For sale by the U.S. Government Printing Office
Published annually ± 600 pages

This is a directory of U.S. Environmental Protection Agency and other public sector environmental information resources. It includes descriptions and contacts for clearinghouses, databases, documents, hotlines, libraries, models, and government records.

American Chemical Society Environmental Buyers' Guide
Luis Gonzalez, Editor
American Chemical Society, Buyer's Guides, Room 210, 1155 16th Street, N.W., Washington, DC 20036
Published annually, free to qualified buyers and specifiers, $50 to libraries

The intention of this guide is to provide those working in environmental research, remediation, or analysis with a comprehensive guide to obtaining supplies, equipment, or services. The 1993 guide has over 1,500 product and service listings from more than 1,600 companies.

Chemical Engineering Equipment Buyer's Guide
(see *Chemical Engineering* for publisher information)

This guide is designed to put you in touch with the makers of all types of equipment, chemicals, and engineering materials used in the chemical process industries, as well as the suppliers of design, engineering, and construction services. The 1994 guide has some 2,000 product categories listed, with entries from over 4,000 firms.

Pollution Engineering Yellow Pages
(see *Pollution Engineering* for publisher information)

This guide presents a comprehensive listing of environmental instrumentation, equipment, supplies, components, materials, and services with alphabetical company listings.

Pollution Equipment News Buyer's Guide (see *Pollution Equipment News* for publisher information)

The 1994 edition of this guide includes listings from over 2,500 suppliers. The key words include major functions, pollutants, classes of equipment, and when appropriate, the product type.

Water Environment Federation Buyer's Guide & Yearbook
Water Environment Federation
601 Wythe Street, Alexandria, VA 22314–1994
Published annually, included with Federation membership

The 1994 edition of this guide includes a listing of more than 200 categories of products and services and the companies that provide them. It also includes brief descriptions of some manufacturer's products.

APPENDIX 1 BOATYARD WASTEWATER TREATMENT EXAMPLE

BACKGROUND

This is an example of a wastewater treatment system implemented by Miller & Miller Boatyard in Seattle, WA. The yard owner developed the system on his own initiative in response to a requirement that his wastewater be treated prior to discharge to the sanitary sewer. He went through the steps of characterizing his wastewater, evaluating alternatives, selecting the best alternative for his situation, and implementing the system. This boatyard was included in the Metro Maritime Industrial Waste Project and some of the information presented below is taken from the project report. The Northwest Marine Trade Association also has included the boatyard's pretreatment system in the documentation they prepared to help the industry comply with regulations.

The yard, located on the Lake Union Ship Canal, is a small facility that performs boat repair and maintenance on recreational boats. It has one crane to lift boats out of the water and place them on land for work.

WASTE CHARACTERIZATION

Wastewater is generated by pressure and handwashing of boat hulls. No bilge water is removed from boats at the site, and therefore, bilge water is not handled. The hull wash wastewater has a turbid appearance from particles suspended in solution. The suspended particles are produced by abrasion of the painted surface. They contain bottom paint and marine growth. Copper is the major contaminant of concern. It is present in the paint as an antifouling agent. The paint may also contain zinc and lead. Tin is regulated by restrictions on the application of tributyltin paints. These paints are not applied or routinely removed at this yard. Regulated organics are not present in significant concentrations unless they are spilled or leak during washing operations. About 60 gal of wastewater are produced from a typical job and about 25 hulls are washed each year. Table 29 presents analytical results for untreated hull wash water.

EVALUATION AND SELECTION OF ALTERNATIVES

Treatment of the hull wash wastewater requires the removal of suspended solids to remove the metals. The treatment options considered included simple

Table 29 Analytical Results for Untreated Hull Wash Water

No.	pH (units)	Suspended solids (mg/L)	As (mg/L)	Cd (mg/L)	Cr (mg/L)	Cu (mg/L)	Ni (mg/L)	Pb (mg/L)	Zn (mg/L)
1	7.4	1280	0.07	<0.004	0.05	95	0.05	3.5	6.5
2	6.7	1400	0.07	0.004	0.16	190	0.09	3.4	6.9
3	7.1	260	<0.05	0.004	0.03	8	<0.01	0.5	1.7
4	7.5	570	<0.05	0.010	0.05	77	0.06	2.2	9.6
5	5.8	—	<0.05	0.006	0.03	27	0.03	1.7	2.5

settling, filtration, and chemical precipitation. The treated water disposal options included discharge to the sanitary sewer and discharge to receiving waters.

To test simple settling, two batches of untreated wastewater were allowed to set undisturbed for 24 h. Samples were poured off and analyzed. Copper concentrations were 23 and 36 mg/L. Zinc concentrations were 5.0 and 2.2 mg/L. These results indicated that simple settling was not effective on the wastewater. It was ruled out as a viable treatment method.

To test filtration, the wastewater was taken to the laboratory and poured through a coarse paper filter. The filtrate contained copper at 1.0 mg/L and zinc at 0.08 mg/L. The wastewater would meet sanitary sewer discharge limits. However, the soft paint particles and biological material in the wastewater rapidly plugged up (blinded) the filter media. This indicated that filtration of raw wastewater was not feasible. It was ruled out as a viable treatment method.

To test chemical precipitation, wastewater was collected in 35-gal containers. The following recipe was used:

- Add 0.5 g/L aluminum sulfate (alum) or about 60 g or 2.5 oz to 30 gal
- Mix for several minutes
- Add about 1 g/L lime to bring the pH to about 8 (about 5 oz to 30 gal)
- Mix for several minutes
- Check pH with pH paper, add more lime if needed, and mix
- Allow to settle for 30 min
- Sample and test treated water

The results of three trials are compared with sewer and receiving discharge limitations in Table 30. The treated water met the sanitary sewer discharge limitations, but not the receiving waters discharge limitations. Based on the successful trials, chemical precipitation was selected as the wastewater treatment method. Discharge to the sanitary sewer was selected as the treated water disposal method because that option was available and the discharge limitation could be met cost effectively.

In several preliminary trials, lime was not added to raise the pH. When no lime was added, the metals were much higher than sanitary sewer discharge limitations, even though the water cleared up and solids settled. This confirmed the need to raise the pH above 8 for effective treatment.

Table 30 Results of Three Treatment Trials Compared to Wastewater Discharge Limitations

		Permit limits		Wastewater conc.		
Parameter	Unit	Sanitary sewer	Receiving water	Untreated (average)	Treated (average)	(high)
pH	U	5.5–12	6.0–8.0	—	8.6	11.6
Suspended solids	mg/L	Not set	45	420	82	110
Arsenic	mg/L	1.0	Not set	<0.05	<0.05	<0.05
Copper	mg/L	3.0	0.0029	42	0.6	0.84
Lead	mg/L	2.0	0.14	1.3	0.06	0.09
Zinc	mg/L	5.0	0.095	5.7	0.16	0.26

Note: The averages here are based on three full-scale wastewater treatment trials. The samples were taken before and after treatment of a batch of wastewater. The untreated average does not correspond to the previous table because different samples were taken.

IMPLEMENTATION OF THE SYSTEM

The first task was to figure out how to collect the wastewater. The boats are washed outside and the wash area drains to the receiving water. To collect the wastewater a large tarp is spread out on the ground and the boat and supporting structure is placed on it. A flexible 4-in. plastic pipe is placed around the edge under the tarp to act as a berm, allowing the water to pool in the tarp. The yard was paved to facilitate use of the tarp and prevent contamination of the soil. The collected wastewater is pumped with a standard sump pump to the treatment containers. The tarp and pump are rinsed with tap water to clean them after use. Figure 14 shows the wastewater collection tarp setup and Figure 15 shows it in use. The tarp cost $500 and the pump, hoses, and fittings were about $600. Figure 16 shows a wastewater collection system in use at an another boatyard.

The wastewater is treated with the above recipe in two 35-gal plastic garbage cans which cost $20 each. After treatment the water is pumped from the garbage cans to the sanitary sewer using the same sump pump used to fill them. Analytical results for a recent batch of wastewater are given in Table 31. The treatment facility, which includes a bermed and covered enclosure, is shown in use in Figure 17. It cost $500.

The settled sludge produced by the treatment process is mostly water and is typically 4% solids. The sludge is accumulated in a third garbage can and periodically dewatered by filtration to reduce its volume. The solids content could be increased to 30 to 40% by removing the water with a filter press. However, a filter press costs at least $3,000 and would have a long payback time because of the small volume of sludge produced. Therefore, gravity filtration with a bag filter is used to reduce the water content. About 1 gal of dewatered sludge is produced from each boat. The sludge can be disposed for $12/gal or about $300/year. It costs more to test the sludge to determine if it is a hazardous waste than it does to assume that it is a hazardous waste and pay

Figure 14 The wastewater collection tarp setup.

Figure 15 The wastewater collection tarp in use.

for disposal as such. The operator takes the sludge to a hazardous waste disposal facility in 5-gal lots.

A wastewater discharge authorization was obtained from the local industrial wastewater pretreatment authority to allow discharge of the wastewater to the sanitary sewer. A discharge permit application was completed to include

Figure 16 A wastewater collection system in another boatyard.

Table 31 Recent Hull Wash Wastewater Treatment Results

Parameter	Units	Treated wastewater conc.
Arsenic (Method 206.3)	mg/L	<0.005
Copper (Method 220.1)	mg/L	0.47
Lead (Method 239.1)	mg/L	0.03
Zinc (Method 289.1)	mg/L	0.028

untreated and treated wastewater data, a description of the collection and treatment process and sludge disposal method, and a diagram of the system similar to the one shown in Figure 18. The wastewater discharge authorization was obtained at no cost because the discharge was small.

It would cost $12/gal or about $720/batch or $18,000/year, assuming 25 jobs, plus transportation and administrative fees to send the wastewater to an off-site facility for treatment. The wastewater would still need to be collected with the tarp and pumped into a container for disposal. This cost estimate assumes that the waste would be treated by the same facility that handles the sludge and that no volume discount would be offered. The cost to send a batch of wastewater offsite for treatment is about three times what the yard charges to haul a boat out of the water and would be cost prohibitive.

The total cost of the treatment system was $1,560. It took about 100 hours to select, design, and build. The tarp takes 30 min to set up and take down. Each batch of wastewater requires 1 hour to treat, for a total of about 40 hours/year. The time it takes to treat the wastewater onsite is comparable to what it would

Figure 17 The wastewater treatment system used by Miller & Miller Boatyard.

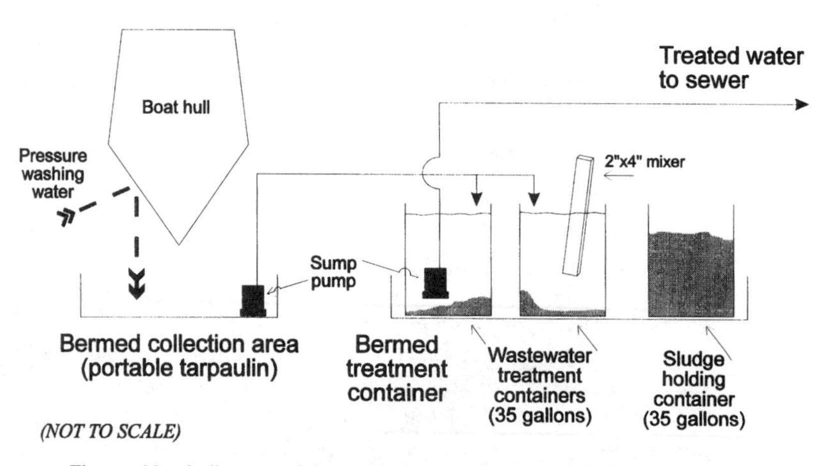

Figure 18 A diagram of the wastewater collection and treatment system.

take to haul it away. The implementation, operation, and sludge disposal costs totaled about $6,000 the first year, assuming a labor rate of $30/hour. Ongoing operation and disposal costs are about $1,500/year or about 10% of the cost of hauling untreated wastewater.

APPENDIX 2 AUTOMOTIVE MACHINE SHOP WASTEWATER TREATMENT

INTRODUCTION

This is an example of an evaporator which was successfully implemented by an automotive machine shop. The shop owner was interviewed to determine the following information:

Reason for putting in the system
Description of the process generating the wastewater
Wastewater characterization — volume, analysis, frequency of generation
How the system was selected, including testing and evaluation
Cost of installation
Operation and maintenance requirements and cost
Regulatory requirements — pretreatment program, PSAPCA, and others
Resultant cost savings

REASON FOR INSTALLING THE SYSTEM

The shop owner developed the system on his own initiative because he wanted his new facility, occupied in 1987, to comply with environmental regulations. He read the *Federal Register* and it was clear to him that his small business was affected by the regulations. He felt that he would be under scrutiny from the pretreatment and hazardous waste authorities. He wanted to deal with his shop's waste responsibly, protect his business from environmental liability, and provide a good working environment for the workers in his shop.

DESCRIPTION OF THE PROCESS GENERATING THE WASTEWATER

Old Facility

A caustic hot tank was used to remove heavy oil and grease contamination from steel parts. Rinsewater and spent hot tank solution were discharged to a floor drain which drained directly to the sanitary sewer. The shop owner knew the caustic was beneficial to the sewer system and was blissfully unaware of the metal contamination. Petroleum distillate was used to clean light residue from steel parts and all of the oil and grease contamination from aluminum parts. Solvent rinsed from the parts was allowed to go to the sewer. Mop water from floor clean up was discharged to the floor drain. No records of wastewater characteristics or volume were retained.

New Facility

A separate room was built to house a parts-cleaning facility. An open flame cleaner is used to burn the oil and grease contamination off steel parts. A media blast cabinet, using steel shot, is used to remove the ash produced by open flame cleaning. The cabinet debris is managed as hazardous waste. A caustic hot tank is used only to clean small steel parts. A hot detergent jet washer is used to rinse parts following the caustic hot tank and to clean aluminum parts which would otherwise be etched in the caustic hot tank, an undesirable effect. Petroleum distillate, used to clean parts in use on the main shop floor, is segregated from the wastewater-generating processes and is sent off-site for reclamation and reuse. The process changes greatly reduced rinsewater and spent hot tank solution generation, as compared to the old facility. The reduced wastewater volume made the use of an evaporator practical.

WASTEWATER CHARACTERIZATION — VOLUME, ANALYSIS, AND FREQUENCY OF GENERATION

Wastewater is produced from the hot tank, jet washer (a pressure spray cabinet), and floor cleanup. Two types of wastewater are generated by hot tank cleaning of automotive engine parts, rinsewater and spent hot tank solution. The contaminants removed from the parts usually cause the hot tank solution and rinsewater to exceed discharge limitations for oil and grease, lead, and zinc. The jet washer solution contains emulsified oil and may contain lead and zinc.

An estimated 1,500 gal/year of wastewater are produced from all of the processes combined. The 1,500 gal/year estimate is based on the 10 to 1 volume reduction observed in the initial evaporation shop trials. No wastewater is discharged to the sewer. All of it, including floor mop water, is put into the evaporator. The evaporated water is vented to the atmosphere outside of the shop. The evaporator produces about three drums per year or about 150 gal/year of sludge. An analysis of the evaporator sludge is presented in Table 32. The pH is >12.5 due to the caustic, and lead exceeds the TCLP limit. The sludge is designated a dangerous waste according to EPA criteria.

HOW THE SYSTEM WAS SELECTED, INCLUDING WHAT TESTING AND EVALUATION WAS DONE

The shop owner previously learned from wastewater analysis that an oil water separator did not remove metals from his wastewater effectively enough

Table 32 Evaporator Sludge Analytical Results

Component	%	Metal (TCLP)	Conc. (mg/L)
Water	5–15	Arsenic	0.87
Oil	10–20	Cadmium	0.17
Sludge/sediment	50–60	Chromium	0.04
Metal gasket debris	1–10	Copper	2.9
Sodium phosphate	1–10	Lead	32
Sodium metasilicate	10–20	Nickel	0.22
Ethylene glycol	<1	Zinc	1.1
Sodium hydroxide	1–5		
pH (units)	>12.5		
Flash point	None		

EPA hazardous waste codes = D002, D008.

to meet discharge limitations. He did not know what kind of treatment system would meet his needs or what was available.

He started his selection process by looking at what was being offered for the automotive machining industry. The only technologies identified included filters and centrifuges. He reviewed what literature he could find and talked to others about these technologies. He determined that filtration and centrifugation appeared too labor intensive to fit well into his operation.

He then contacted the marine industry because he knew it generated wastewater similar to that of his shop in some processes. He found an electrochemical process which was claimed to remove contamination from wastewater through coagulation with electrically charged plates. He obtained a pilot unit and ran it in his shop. The process did not work on his wastewater. The oil and dirt in the wastewater collected on the plates, became slimy, and plugged the unit.

He obtained treatment chemicals from a vendor at a trade show and tried to treat his wastewater in a 55-gal drum following the recipe provided by the vendor. He had the treated water analyzed at a contract laboratory and found that the lead concentration was still over the discharge limitation of 4 mg/L. He called the vendor and reported the results. After assuring him that this problem had not arisen before, the vendor was never heard from again. The shop owner ruled out chemical precipitation.

Chemical precipitation treatability test results for hot tank rinsewater from another automotive machine shop are presented in Table 33. The bench test data indicate that the rinsewater is alkaline, equivalent to about 0.1% NaOH, based on the amount of acid needed to neutralize it. The oil and metal are present as finely divided particles which do not settle readily. Simple neutralization does not result in effective metal removal. To remove the metal it is necessary to drop the pH to about 3 to allow the particles in the wastewater to destabilize and aggregate. Then, when the pH is raised to about 8.5, the metals and oil effectively precipitate. Enough metal, such as iron, is in the solution already to aid in flocculation. Anionic polymer addition aids the precipitation

Table 33 Chemical Precipitation Treatability Test Results for Hot Tank Rinsewater

Sample	Contaminant conc. Pb (mg/L)	Zn (mg/L)	FOG (mg/L)	Description of test procedures and appearance of samples
Untreated	23	24	730	Dark turbid water, no perceptible floating oil
Test 1	15	17	54	500 ml, pH = 11.9, + 1 ml $FeCl_3$ solution, pH = 11.5, + 0.8 ml 35% HNO_3, pH = 8.3, + anionic polymer, poor settling, dark turbid water
Test 2	<0.025	0.18	25	500 ml, pH = 12.1, + 1.5 ml 35% HNO_3, pH = 3.1, particulate formed, tends to float, + lime to pH = 8.5, small floc, settled well, water slightly turbid, + anionic polymer, bigger floc, good settling, clear brown-tinted water

process. Chemical precipitation requires the use of concentrated acid and more operator involvement than evaporation.

The shop owner inquired about an off-the-shelf ultrafiltration system. An ultrafilter would remove the oil contamination, but would not remove metals effectively. He made the decision to rule out ultrafiltration, based on what he learned.

The shop owner obtained information on an evaporator which was developed in Southern California. The vendor told him that the trend in Southern California was to eliminate discharge to the sanitary sewer because of the stringent discharge limitations. He worked with a local representative to install the unit in his shop. He had worked with this representative in the past to install process equipment, including the open flame cleaner, knew the representative was reliable, and trusted the representative to get the evaporator working effectively. He bought the evaporator and the representative helped to get it running. The representative worked with the Washington State Department of Ecology (WDOE), which regulates hazardous waste in Washington State, and the Puget Sound Air Pollution Control Agency (PSAPCA) to convince them to accept evaporation as an environmentally acceptable treatment option. The evaporator has been in use in his shop for about 5 years.

Treatment required the removal of oil and metals. The treatment options considered included oil water separation, ultrafiltration, chemical precipitation, electrochemical precipitation, and evaporation. The treated water disposal options included discharge to the sanitary sewer and venting to the atmosphere. The selection of the waste treatment residue disposal option depended on whether it was designated a hazardous waste. Evaporation was selected as the most viable alternative, with venting of the evaporated water and disposal of the sludge as hazardous waste.

Figure 19 The evaporator (left) receives solution from the hot tank (right) and from the jet washer, shown in Figure 20.

COST OF INSTALLATION

The 5 gal/hour evaporator cost about \$3,000 and installation cost about \$500. Figure 19 shows the evaporator and hot tank in use. The evaporator took about 100 hours to select, install, and get running. An air pollution control permit was not required for the unit. The total cost of implementing the treatment system was \$9,000. The jet washer (shown in Figure 20) is used to rinse parts following the hot tank. Jet washer water is periodically pumped to the evaporator. The open flame cleaner and media blasting cabinet (shown in Figure 21), used to reduce the use of the caustic hot tank and subsequent wastewater generation, cost about \$30,000 and about \$2,000 to install. A notice of construction for the flame cleaner was filed, at no cost, with the local air pollution control authority.

The shop is working to install an oil skimmer in the evaporator to remove oil from the top of its contents. The oil layer restricts the evaporation of water, thus reducing efficiency. The skimmer will be operated at night when the evaporator is shut down.

OPERATION AND MAINTENANCE REQUIREMENTS AND COST

The evaporator runs unattended except for sludge removal. Sludge is shoveled from the evaporator about every 2 weeks. It takes about 1 hour to remove and drum the sludge, a total of about 25 hours/year. At a rate of \$25/hour, operational labor costs about \$600/year.

Figure 20 Solution from the jet washer (left) is pumped to the evaporator (right).

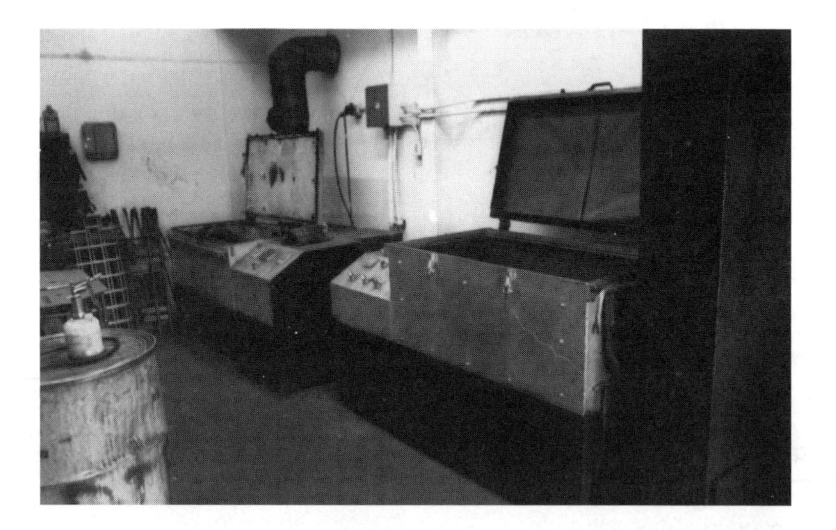

Figure 21 The flame cleaner, used to burn contaminants from parts, is shown on the left. The media blast cabinet, used to remove residue from the parts following flame cleaning, is shown on the right.

The sludge is accumulated in a drum. Periodically water is removed from the top of the sludge and returned to the evaporator to reduce the volume of sludge disposed. The sludge is taken to a hazardous waste disposal facility in 55-gal lots. The disposal cost for each drum is currently $453, including pickup and transportation, plus $50 for the drum itself. Sludge disposal costs about $1,500/year.

The evaporator draws 9600 W for 8 hours/day, 5 days/week, about 1500 kWh/month. At $0.04/kWh, the electricity cost is about $60/month or $700/year.

In 5 years the bottom of the evaporator had to be replaced three times at a cost of about $2,000 per replacement. The electrical heating elements had to be replaced three times at a cost of about $1,000 per replacement. Therefore, a total of about $9,000, or $1,800/year, was spent on evaporator repairs.

The total operation and maintenance costs are about $4,600/year.

REGULATORY REQUIREMENTS — PRETREATMENT PROGRAM, PSAPCA, AND OTHERS

The shop owner submitted process, waste volume, and waste characterization information to the local pretreatment authority. The local industrial wastewater pretreatment authority reviewed the facility and issued a discharge authorization letter, at no cost, which included discharge limitations. The shop is not subject to the discharge authorization requirements because no wastewater is discharged. The discharge authorization was obtained to document that the pretreatment authority had been contacted and their requirements are understood. No formal wastewater discharge permit was required. A permit, which cost $1,500, would require periodic sampling and inspections by the pretreatment authority.

Evaporation is legal, in Washington State, as a treatment-by-generator option. A hazardous waste treatment facility permit is not required.

The shop owner obtained a hazardous waste generator number from the state, maintains waste manifest records, and files yearly waste reports. He has chosen to follow the regulatory requirements of a regulated hazardous waste generator rather than managing his waste as a nonregulated generator in order to fully document his waste management practices. He chose to have a licensed hauler pick up his sludge in 55-gal lots rather than taking it to the facility in 5-gal buckets, to minimize his liability and circumvent the inconvenience of making monthly trips to the treatment facility.

An air pollution control permit is not required by the local air pollution control agency because water vapor is not considered an air pollutant unless it is creating a nuisance.

RESULTANT COST SAVINGS

If he chose to do nothing about his wastewater and not comply with the wastewater discharge regulations, he could have saved the time it took to find the system, the $9,000 cost of installation, the $4,600/year to evaporate the water and dispose of the sludge, and the $30,000 for the flame cleaner, but his business would be in jeopardy. He chose to take care of the problem to reduce his liability and to maintain the reputation of his business in the industry.

Noncompliance was not an option for this business owner. The cost savings of the system is therefore compared to the cost of drumming up and shipping all of the wastewater off-site for treatment. About 30 drums per year of wastewater are generated. It would cost $500 per drum or about $15,000/year to dispose of the wastewater. The wastewater would still need to be collected and pumped into containers for disposal. Wastewater handling and recordkeeping requirements are similar for both alternatives.

The system cost $9,000 to implement. Yearly operation, maintenance, and disposal costs are about $4,600. On this basis, the treatment system paid for itself in less than 1 year.

APPENDIX 3 OILY WASTEWATER TREATMENT EXAMPLE

INTRODUCTION

This batch chemical treatment system was designed to treat oily water decant from 2,000-gal eductor trucks. The trucks vacuum water from oil water separators installed in parking lots and industrial maintenance sites. The system uses alum and lime to precipitate oil and metals. The treated water is discharged to the sanitary sewer.

The vactor trucks originally traveled directly to an off-site waste treatment facility to discharge their loads. In time, the cost of disposal was increased from $0.50 to $1.25/gal, causing the yearly disposal cost for 100,000 gal of wastewater to increase from $50,000 to $125,000/year. This provided the impetus to develop an on-site treatment system to reduce the volume of wastewater sent off-site for disposal.

The loads typically consist of 70% free liquid and 30% settled solids. The liquid contains between 200 and 500 mg/L fat, oil and grease (FOG). The sanitary sewer FOG discharge limitation is 100 mg/L. Zinc is present in the wastewater, but it is typically under the sanitary sewer discharge limitation of 10 mg/L.

TREATMENT SYSTEM DEVELOPMENT

The site environmental project manager knew an engineer in another section of the organization who had designed wastewater pretreatment systems. She asked for his guidance to help implement a treatment system. The engineer spent about 25 hours recommending bench and pilot test procedures and minimum treatment system design requirements. The engineer knew from previous experience in similar situations that chemical precipitation would produce wastewater that would meet discharge limitations and be cost effective in this situation.

The first stage of the treatment process involved pumping the wastewater out of the vactor trucks, leaving the solids behind (Figure 22). This procedure was developed by the project manager and the vactor truck operators. The truck tank is used as a settling chamber to allow the solids to settle overnight. The solids level in a truck is then measured and a PVC stinger pipe, attached to a 1.5-in. hose, is inserted to within 4 to 6 in. of the solids. A diaphragm pump is used to pump the wastewater to an existing 5,500-gal underground vault. The

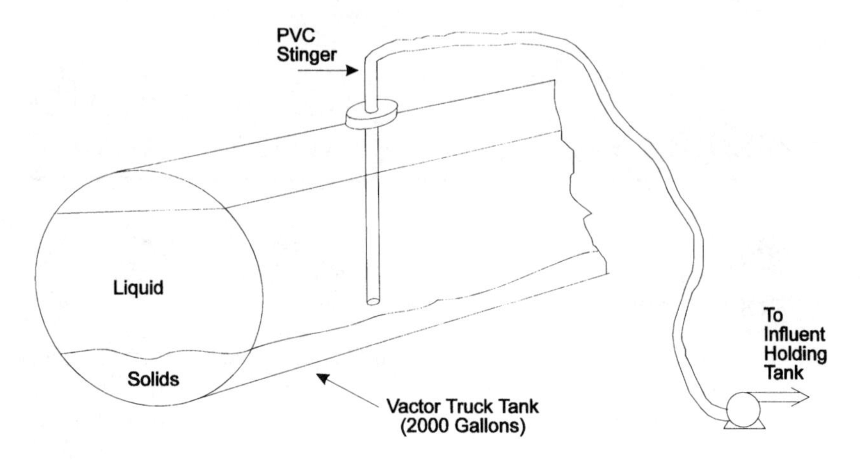

Figure 22 The first stage of the oily wastewater treatment process, decanting from the vactor truck.

solids remaining in the trucks are hauled to the off-site treatment facility for dewatering. Future plans include investigation of on-site dewatering of those solids to further reduce disposal costs.

Wastewater is pumped from the underground holding tank to the 1,400-gal batch treatment system when >3,600 gal of wastewater has been collected. A 192-gal/min pump and a 2-in. hose are employed. The hose is placed below the floating oil that accumulates in the tank, enabling the tank to function as an oil water separator. This keeps floating oil out of the chemical treatment tank. The floating oil is periodically vacuumed out by a company that recovers the oil.

Wastewater samples were taken from the vault, and bench and pilot tests were conducted to confirm that chemical precipitation would be effective prior to building the system. The pilot testing is described in Chapter 5.

The treatment system tank volume was selected to provide for the treatment of one or two batches of wastewater per week. This frequency of treatment could be worked into the vactor truck operators' schedules. They also operate the treatment system. Additional treatment capacity is available if needed by running more batches per week.

The treatment system was built by facility staff onsite. A process flow diagram (Figure 23) and feedback from the project manager was sufficient for the staff to design and construct the system shown in Figure 24. It was built using as much available equipment as possible. A system components list is given in Table 34. It is skid mounted and portable, so that no building modifications were needed to install it. It is operated by compressed air, so that the only utility hookups required were water and compressed air. Not using electricity simplified the installation.

The project manager learned enough about the treatment process during the bench and pilot testing that she was able to start up the system on her own. She involved the operators from the beginning of the project and they

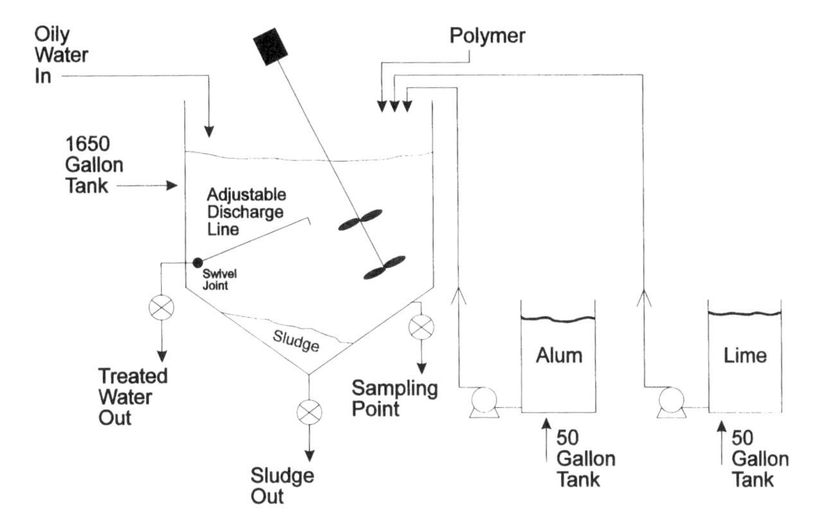

Figure 23 The oily wastewater treatment system.

enthusiastically learned to run the process. Treated wastewater and sludge test results are presented in Table 35. The results document that the treated water meets discharge limitations and that the sludge is not hazardous waste.

OPERATIONS AND MAINTENANCE

The project manager prepared a detailed, step-by-step description of operations and maintenance procedures. A troubleshooting section was included to provide possible solutions to the most common problems encountered by the operators. An outline of the procedures is provided here to suggest what should be documented to ensure that a treatment system is run well. The details will vary for different situations depending on the nature of the setup and equipment. Be sure to include specific instructions for pump, valve, mixer, and other component operation and maintenance procedures.

1. Decant Procedures
   - Measure the sludge depth and liquid level in the vactor truck
   - Insert the stinger to within 4 to 6 in. of the sludge
   - Pump the wastewater into the holding tank; floating oil may also be included
   - Clean the hoses by pumping fresh water through them
2. Influent Pumping Procedures
   - Measure the wastewater level in the holding tank
   - Pump 1,400 gal to the treatment tank, provided at least 42 in. are in the holding tank
   - Do not pump floating oil to the treatment tank
   - Drain and store the hoses and pump

Figure 24 The oily wastewater batch treatment system.

3. Batch Treatment Procedures
   - Turn on the main tank mixer and the two treatment chemical tank mixers, increase speed until the liquid vortexes
   - Pump 7 gal of alum solution into the treatment tank; follow the alum with about 1 gal of water to clear the line
   - Mix the treatment tank for a minimum of 5 min
   - Take a wastewater sample and visually check to determine if small floc particles have formed; if not, see troubleshooting section
   - Pump 4 to 5 in. of lime solution into the treatment tank and mix
   - Take a wastewater sample and check the pH; if <8.0, add small amounts of lime until the pH is between 8 and 11

Table 34 Oily Wastewater Treatment System Components

Component	Cost
Vactor truck decant system	
Sandpiper™ 1½ in. air diaphragm pump, model SA1-A	Free (worth ~$600)
1½ in. reinforced hose with cam lock fittings	$100
PVC stinger	$10
Wastewater transfer system	
Homelight™ 2-in. 192 gal/min gasoline centrifugal pump, model 130232	$600
2-in. reinforced hose with cam lock fittings	$150
Treatment tank	
1650 gal Cal Poly™ open top conical tank with basket stand and mixer mount	$4000
Grovac, Inc.™ air propeller mixer, model 1303	Free (worth ~$1500)
2-in. PVC pipe, valves, fittings, and bulkhead fittings	$500
¾ in. compressed air lines with fittings	Unknown
1 in. portable water hose	$30
Chemical feed system	
Two 50- to 55-gal chemical solution tanks	$160 each
Two Gast™ air propeller mixers, model 2AM-FCW	$350 each
Two Marathon™ ½-in. air diaphragm pumps, model MP02P	$325 each
1 in. PVC pipe, valves, and fittings for treatment chemical transfer	$100

Table 35 Treated Oily Wastewater and Sludge Analytical Results

Sample	Parameter	Conc. (mg/L)
Treated wastewater		
#1	FOG	11
#2	FOG	17
#3	FOG	13
#4	FOG	13
#5	FOG	12
Sludge #1	Cadmium (TCLP)	0.016
	Chromium (TCLP)	<0.01
	Lead (TCLP)	<0.05
	Nickel (TCLP)	0.04
	Zinc (TCLP)	0.05
Sludge #2	Cadmium (TCLP)	0.011
	Chromium (TCLP)	<0.01
	Copper (TCLP)	<0.025
	Lead (TCLP)	<0.05
	Nickel (TCLP)	<0.04
	Zinc (TCLP)	0.02

3. Batch Treatment Procedures (continued)
   - Rinse the lime pump and hose after the pH is adjusted
   - Turn off the chemical tank mixers
   - Mix the treatment tank for 5 to 10 min, turn mixer down to medium speed
   - Add 1 L of polymer solution to the wastewater by pouring into the vortex
   - Visually inspect the floc particles; if not growing, add an additional 0.5 to 1 L of polymer solution

3. Batch Treatment Procedures (continued)
    - Mix for 5 min, collect a sample and allow the floc to settle; the top layer should be clear
    - Repeat the treatment procedure if needed
    - Turn off the mixer; allow the tank to settle overnight
    - Discharge the clear supernatant to the effluent-holding tank using the adjustable discharge arm, leaving the sludge in the tank
4. Sewer Discharge Procedures
    - Discharge the effluent holding tank to the sewer after three to four batches have been collected
    - Measure the treated water depth and calculate the volume
    - Obtain samples to test for pH, total metals, and FOG
    - Gravity-drain treated water to the sewer
    - Close the drain valve when finished
5. Sludge Disposal Procedures
    - Remove sludge from the treatment tank when about 200 gal of it collects
    - Connect a vactor truck vacuum hose to the 6-in. cam lock fitting on the tank and educt the sludge into the vactor truck
    - Dispose of the sludge along with the untreated solids remaining in the vactor truck after the wastewater has been decanted
6. Oil Disposal Procedures
    - Allow waste oil to accumulate in the vault until the floating layer is 3- to 4-in. thick
    - Call the oil hauler and ask for a pick up; they will skim the oil using a vacuum truck
    - Clean out the vault periodically when necessary
7. Chemical Mixing Procedures
    - Wear the proper safety equipment (coveralls, dust mask, goggles, and gloves) when mixing the chemical solutions
    - Alum and lime
        - Fill the alum or lime solution tank 50% with water
        - Turn on the mixer; increase the speed until a vortex forms
        - Add 50 lb of alum or lime
        - Bring the water level up to the 50-gal mark
    - Polymer
        - Obtain the polymer (Magnifloc™ 1883A Flocculant) in 5-gal buckets
        - Mix 250 ml in 5 gal of water to prepare a working strength solution
        - Clean up any polymer spills with towels, not water, to avoid a dangerous, slippery mess
8. Paperwork
    - Batch treatment log
        - Fill out a complete entry on the treatment log sheet
        - Include the date, amount of chemicals added, total volume treated, amount of sludge, volume left in the tank, volume discharged to the effluent holding tank, pH of the effluent, and the operator's initials
    - Sewer district log
        - Total the volume discharged to the sewer at the beginning of each month

Fill out a discharge notification form and attach a copy of the treatment log

Send the form to the sewer district billing office

Self-monitoring forms

Complete an industrial waste self-monitoring form monthly

Include any effluent holding tank sample results

Include the date and volume of all sewer discharges, and the date and volume of all batch treatments

REGULATORY REQUIREMENTS

Wastewater characterization data, pilot test data, and a description and diagram of the pretreatment system was submitted to the pretreatment authority in a discharge permit application. A wastewater discharge authorization was obtained from the local pretreatment authority at no cost because of the small volume of wastewater. It includes wastewater discharge limitations and provides for self-monitoring to demonstrate compliance.

The wastewater, solids, and sludge produced in the process were tested and determined not to be hazardous waste. Therefore, hazardous waste regulations do not apply to the treatment system. No flammable or combustible materials are used in the treatment process. Therefore, the fire department required no permit for the system. The local air pollution control authority did not require a permit because the treatment system is not removing volatile organics or other regulated air pollutants. The treatment system did not require a building permit. Washington State Industrial Safety and Health Administration regulations must be followed.

RESULTANT COST SAVINGS

Over 70,000 gal of wastewater were treated and discharged in the first year of operation. Sending the wastewater off-site for treatment and disposal would have cost about $87,500. It takes about 1 hour to treat a 1200-gal batch or about 70 hours for the 70,000 gal. Therefore, the labor cost was about $1,750, at a rate of $25/hour. Building the treatment system cost $6,500 for parts, plus 100 hours of the craftsman's time (about $2,500 at $25/hour), 80 hours of the project manager's time (about $2,400 at $30/hour), and 25 hours of the engineer's time (about $900 at $35/hour). The total cost of the installation was about $12,500. The treatment system quickly paid for itself. The operators take pride in treating the wastewater onsite, helping to reduce the cost of doing business for their organization.

Index

G

H

I

T